Dedication

To Claudette and Gérard Beaulieu, and to all who raise critical thinkers.

Anne Beaulieu & Sabina Leonelli

DATA
AND
SOCIETY

A Critical Introduction

Los Angeles | London | New Delhi
Singapore | Washington DC | Melbourne

Los Angeles | London | New Delhi
Singapore | Washington DC | Melbourne

SAGE Publications Ltd
1 Oliver's Yard
55 City Road
London EC1Y 1SP

SAGE Publications Inc.
2455 Teller Road
Thousand Oaks, California 91320

SAGE Publications India Pvt Ltd
B 1/I 1 Mohan Cooperative Industrial Area
Mathura Road
New Delhi 110 044

SAGE Publications Asia-Pacific Pte Ltd
3 Church Street
#10-04 Samsung Hub
Singapore 049483

Editor: Alana Clogan
Assistant editor: Ozlem Merakli
Production editor: Manmeet Kaur Tura
Copyeditor: Camille Bramall
Proofreader: Jill Birch
Indexer: Cathryn Pritchard
Marketing manager: George Kimble
Cover design: Francis Kenney
Typeset by: C&M Digitals (P) Ltd, Chennai, India

Library of Congress Control Number: 2021935490

British Library Cataloguing in Publication data

A catalogue record for this book is available from the British Library

ISBN 978-1-5297-3254-2
ISBN 978-1-5297-3253-5 (pbk)

CONTENTS

LIST OF FIGURES AND TABLE

Figures

Table

DATA STORIES

ABOUT THE AUTHORS

Anne Beaulieu is Professor of Knowledge Infrastructures and Director of the Data Research Centre, University of Groningen. She holds the Aletta Jacobs Chair of Knowledge Infrastructures at Campus Fryslân and developed the minor programme Data Wise: Data Science in Society with Gert Stulp. Beaulieu's research focuses on diversity and complexity in knowledge infrastructures, with particular attention to digital data assemblages and interfaces for the creation and circulation of knowledge. She is co-author of *Virtual Knowledge: Experimenting in the Humanities and Social Sciences* and has published widely on the significance of ethnographic methods for the study of data practices. Beaulieu was Visiting Research Fellow at the Pufendorf Institute for Advance Studies, Lund University, Sweden (2017–2018) and Visiting Professor at the Science and Technology Studies Department of the University of Vienna, Austria (2018). Since September 2018, she has been co-coordinator of the PhD training network of the Netherlands Graduate Research School of Science, Technology and Modern Culture (WTMC), and has developed numerous innovative courses in the major programme Responsible Planet at University College Fryslân.

Sabina Leonelli is Professor of Philosophy and History of Science at the University of Exeter, where she co-directs the Centre for the Study of the Life Sciences and leads the governance strand of the Institute for Data Science and Artificial Intelligence. Her interdisciplinary research and collaborations focus on the epistemology and governance of scientific data and models, and the role of open science in the global – and highly unequal – research landscape. She is Fellow of the Alan Turing Institute, Academia Europaea and the Académie Internationale de Philosophie de la Science; Editor-in-Chief of *History and Philosophy of the Life Sciences*; and Associate Editor of the *Harvard Data Science Review*. She has served as expert adviser for national and international agencies and received funding from several public funders including two awards from the European Research Council. Her books include the award-winning *Data-Centric Biology: A Philosophical Study* (Chicago University Press, 2016) and *Data Journeys in the Sciences* (Springer, 2020, with Niccolo Tempini).

ACKNOWLEDGEMENTS

This book builds on our work on data from the past two decades. It draws on material from several research projects as well as countless interactions with generous colleagues, collaborators and audiences whose insights and engagement have informed and inspired us.

Many students from around the world contributed to our thinking about data through their astute questions, diverse insights and perspectives, and lively in-class discussions. They include the Master's students in the courses Big Data and Energy (University of Vienna) and Big Data in a Digital Society (University of Groningen), and especially the students in Introduction to Data and Data as Evidence (University of Groningen) and the postgraduate module Data Ethics and Governance at the University of Exeter.

A number of colleagues also acted as sparring partners and we are grateful for their contribution to sharpening the concepts in this book and for inspiring us through their own research. In The Netherlands, thanks go expressly to Ludo Waltman, Salome Scholtens, as well as Oskar Gstrein, Daniel Feitosa, Andrej Zwitter and all other members of the Data Research Centre. In Exeter, thanks go to colleagues at the Exeter Centre for the Study of the Life Sciences (Egenis) and the Institute for Data Science and Artificial Intelligence (IDSAI), especially the wonderful participants in Data Crunch meetings where the book project was discussed; Lora Fleming, Gavin Shaddick, Alberto Arribas and Hywel Williams, who are making the data world a better place; John Dupré, Rachel Ankeny, Gail Davies and Brian Rappert, who unfailingly provided inspiration and support; and the brilliant PhD students at the Environmental Intelligence Centre for Doctoral Training.

As a project, this book benefited from the generous help of Andrej Zwitter, who established a fruitful link with Sage. Sage editors Robert Rojek and Natalie Aguilera provided prompt and helpful steering for the project throughout, and procured six referee reports that contained extremely useful, timely and constructive feedback – many thanks to these referees too. Trish Nowak provided excellent editorial assistance in the final weeks of this project. Other colleagues provided valuable feedback on different chapters that improved the structure and clarity of the material. We want to acknowledge the contributions of Arthur Vandervoort, Marthe Stevens, Clarisse Kraamwinkel, Malcolm Campbell-Verduyn and Esther Hoorn, long-time collaborator on getting things done with data. Selen Eren and Joonas Lindeman were supportive early readers.

In an academic climate that values research activities, this book required us to make a big commitment to teaching – a commitment that was also fuelled by the work of others. In Groningen, we would like to note the contributions of Jasper Knoester and Gerard Renardel de Lavalette for championing a course on Big Data and Society in the Master's in Computer Science; of the many colleagues who helped build the minor programme on data, especially Ronald Stolk, for spearheading what became Data Wise: Data Science in Society; René Veenstra, who helps steer this complex programme; and Gert Stulp, who, in all ways, is an invaluable colleague and the ideal programme co-coordinator. In Exeter, we thank the College of Social Sciences and International Studies, the Department of Sociology, Philosophy and Anthropology, and the Institute for Data Studies and Artificial Intelligence – especially its director Richard Everson – for recognising the centrality of this work to data science training and supporting the expansion of our research and teaching activities. We are grateful to Chee Wong, Jill Williams and Katie Finch for their vital administrative support of data studies activities at the university through Egenis and IDSAI.

Inspiration for our teaching has come from many corners and we are lucky to be surrounded by colleagues committed to interdisciplinary teaching, to learning for both students and lecturers, and to creating optimal conditions for students' development. Anne would especially like to thank Malvina Nissim, Sepideh Yousefzadeh, Elena Cavagnaro, Berfu Unal, Tassos Sarampalis, Indira van der Zande, Ineke Visser, Amaranta Luna Arteaga, Engelien Reitsma, Piet Bouma, Bernike Pasveer, Nishant Shah and Hanny Elzinga.

Finally, the main parts of this book were written during the COVID-19 pandemic. This created considerable challenges to completing the manuscript, including health issues and bereavement for us both, and we would like to thank our husbands and children, Maarten Derksen and Félix Derksen and Michel, Leonardo and Luna Durinx, for keeping us (relatively) sane through it all. Anne would also like to thank the Post-pandemic University, participants to Wednesday Wine, the informal 'coalition of the willing' at the University of Groningen and Jacob Veenstra; while Sabina thanks her closest friends and colleagues for all their support, strength and unfailing sense of humour.

A few passages in this book draw from the following publications:

Beaulieu, A. (2021). Data practices and SDGs: Organising knowledge for sustainable futures. In M. Hojer Bruun, D. Brogaard Kristensen, R. Douglas-Jones, C. Hasse, K. Høyer, B. R. Winthereik and A. Wahlberg (eds), *Handbook for the Anthropology of Technology*. London: Palgrave Macmillan.

Gstrein, O. and Beaulieu, A. (under review). What makes data personal? A presentation of multiple conceptions of privacy in the digital age.

Leonelli, S. (2017). Biomedical knowledge production in the age of Big Data. Report for the Swiss Science and Innovation Council, published online November 2017: www.swir.ch/images/stories/pdf/en/Exploratory_study_2_2017_Big_Data_SSIC_EN.pdf (accessed 23 April 2021).

Leonelli, S. (2018). *La Ricerca Scientifica Nell'Era Dei Big Data*. Rome: Meltemi Editore.

Leonelli, S. (2021). Data science in times of pan(dem)ic. *Harvard Data Science Review*, 3(1). https://doi.org/10.1162/99608f92.fbb1bdd6

Leonelli, S., Lovell, B., Fleming, L., Wheeler, B. and Williams, H. (2021). From FAIR data to fair data use: Methodological data fairness in health-related social media research. *Big Data and Society*, 8(1). https://doi.org/10.1177/20539517211010310

OVERVIEW OF THE BOOK

In today's digital society, a critical understanding of data is essential for all. Knowledge about data is often split into areas of expertise, so that processes that span algorithms, servers, users and institutions are rarely discussed coherently and accessibly. *Data and Society: A Critical Introduction* presents a set of concepts to assess how data shape science, policy and politics, including how data are turned into metrics that are used to make decisions. It connects data as a highly technological practice to broad social questions of evidence, innovation and knowledge.

The book provides an analytical framing to understand the role of data in contemporary society and foster good data practices. Our analysis is grounded on the following three ideas:

1. Data are not an autonomous force or a unidimensional technical fix, and the use, valuation, circulation and deployment of data are shaped by social and material factors, including social institutions and technologies.
2. Nearly all areas of professional and academic work involve interactions with data science, and the skills to relate, evaluate and shape data practices are necessary to be able to exercise one's expertise responsibly.
3. The ability to write code and develop algorithms contributes but is not sufficient to understanding and critically assessing data practices. Another essential component of training in data use involves the conceptual and methodological toolkit developed within the social studies of science broadly conceived (including critical data studies, data ethics and the history, philosophy and science and technology studies).

Building on these key insights, the book addresses the growing attention to the social embedding of data across different settings, from business to policy and government, from sports to health and climate change. It explains the challenges that such embedding brings both for the governance data flows and for the technical management and use of data.

This book is intended as an interdisciplinary introductory textbook for advanced undergraduates or graduate students, which connects the phenomenon of datafication and related technologies to social, technological and economic change. Its conceptual framework relates ideas and principles with concrete cases, to help illustrate the growing importance of data in different spheres of knowledge production and its implications for a wide variety of sectors. To this aim, the book is structured around four sets of practices around data, with a series of *data stories* used to discuss salient concepts to concrete issues. The data stories present details about a specific use of data and ask questions about different aspects and implications of that case. In doing so, they exemplify ways to question and scrutinise the broader social implications of data work and highlight how technical aspects of data practices are entwined with institutions, users, regulations, business models and cultural norms. You are invited to read the data stories and think about the questions that they pose in two stages: once before reading the related section of the book, and a second time after having read the materials in the section, which will help you to articulate your own answers to the questions being raised.

At the end of each chapter of the book you will find a list of four additional readings. These are scholarly sources that will be very helpful to you if you want to deepen your understanding of the issues discussed in the chapter. We chose these sources because of their breadth and their accessibility: they are by no means the only key resources available for the topic (as you will realise when you look at the references for each chapter, and run a bibliographic search for the topic). The scholarly literature on data in society is growing exponentially as we write, and there will no doubt be many other useful readings for you to discover by the time you have read this book. We strongly encourage you to browse the available literature on the issues that most interest you, and explore some of the key journals in this field, such as *Big Data and Society* or the *Harvard Data Science Review*.

INTRODUCTION

Let us start with an everyday story common to the world of internet users. It is April 2021 in the United Kingdom. Thirty-year-old Lara is a healthy and active individual. In the evening she occasionally watches Amazon Prime shows, with a strong preference for science fiction series – but generally prefers spending her free time talking to friends and doing sports. Her normal routine is disrupted when she suddenly falls ill and has to stay in bed for two weeks to recover. During that time, she is in such bad condition (headache, fatigue, mental confusion) that all she can manage to watch are costume dramas and teenage romance films, whose pace is slower and whose soundtrack is less jarring for her headache. Once she is back in shape and able to return to her preferred lifestyle, Lara notices that the list of movies recommended to her has changed, and she is finding it harder to identify series that she might like to watch. Other parts of her Amazon account have changed too, with insistent advertising for clothes and products that she dislikes. Her Google account also seems affected by the changes, with adverts for Jane Austen-themed holidays and romantic getaways popping up every time she scrolls down. Lara is upset because the internet platforms she uses for online shopping and entertainment no longer reflect her preferences, and what used to feel like useful shopping tips now feel like useless, intrusive clutter in her online space. Lara is also upset because she had not realised how extensively the information concerning what she watches on Prime would travel to other platforms and online services.

This everyday situation raises many questions. Is Lara right to feel upset in this situation? Is there a problem here, and what is it? Do Lara's watching preferences constitute sensitive and/or personal data? Are these preferences valuable data, and for whom? What can or should Lara do about the mismatch she experiences? Is this situation legal? Is it a necessary condition for the provision of the streaming service? Is it right that Lara should have to deal with this? Is it possible to 'fix' this situation by improving the system's responsiveness to Lara's change in preferences, so that it can be updated more quickly to reflect her changing circumstances? Or should Lara simply adapt her behaviour to the system, so that she does not get caught up in this way again in the future?

Let us now consider a different story, which happened in the early 2010s to an American scholar called Mary Ebeling – who then went on to write an important book about her experience (Ebeling, 2016). After years of attempts,

Mary was delighted to be pregnant and could not wait for the birth of her baby. Tragically, however, during the eighth week of her pregnancy she suffered a miscarriage and lost the baby. As with so many other mothers-to-be finding themselves in that horrible situation, she came home from the hospital distraught, only to find an advertisement on her doorstep that was specifically targeted to pregnant women. Despite her complaints to the companies responsible for the unwanted correspondence, her letterbox continued to fill with advertising and samples from baby product companies. On the week in which the baby should have been born, she even received congratulation notes complimenting her on the birth of a healthy child. The advertising campaign continued for years, tracking the milestones and celebrations that her daughter would have enjoyed had she survived. Mary was devastated by these constant reminders of the baby she had lost and kept pleading with the companies responsible to stop sending her these unwanted gifts – to no avail. Exasperated, Mary did some research and discovered that while she was pregnant, the hospital sold her personal data to a private company as part of a clinical trial that she had agreed to participate in, as a way to get support for her pregnancy. The company running the trial in turn sold Mary's data to a data trading company, which sold them off again to a number of baby product companies. The data were never updated with news of the premature end of the pregnancy, which is why she kept receiving the merchandise. In her initial attempts to understand what had happened to her data, Mary was unsuccessful as companies did not want to release information about what data they own, and whom they acquired the data from. Mary managed to elicit that information only when she revealed her ordeal as a bereaved mother.

This is clearly a situation where something went very wrong in the handling of sensitive medical data. Was the hospital justified in selling Mary's data in the first place? It turned out that she did give her consent to participate in the clinical trial and she signed an agreement that enabled the company to sell her data to third parties. Her participation in the trial may have been beneficial to the development of medical research, which was the reason why she signed up. Does this make the other ways in which her data were used okay? Is participation in medical research equivalent to consenting to data being used for advertising? Are these two things necessarily related? Can we still distinguish research from the commodification of reproduction? What does this case tell us about the entwinement of medical care and business in the American context and the commercialisation of health? Could such a situation be avoided and, if so, how?

Let us now move to a third story, which could well be described as a data triumph. For many years, agronomic institutions around the world have worked on improving systems for sharing data about plant pathogens and diseases, so as to be able to map the spread of new pests and improve understanding of how to treat the affected crops. The decrease in disease in crops has a direct impact on hunger and malnutrition. In 2019, a consortium of public and private research institutions released an app called PlantDisease. This app could be downloaded for free on the smartphones of farmers around the world and helped them to identify diseases in their crops in a timely manner, as well as providing tips on how to treat those diseases. By providing such important information to farmers in remote areas of the globe, the app is helping to boost food production worldwide as well as supporting the livelihoods of farming communities.

We can think of this as a case of data helping to feed the world and this seems the best of what data can do for humanity. Is there any possible drawback to such a development? What happens if the information provided by the app is unreliable or faulty – who takes responsibility for the effects? Also, is the information provided though this app simply a series of 'facts' about plant disease? There are actually many different approaches to understanding and treating plant diseases within agronomic research. Some privilege technology-driven interventions, like using targeted pesticides or genetic modification of seeds. Others favour environmental interventions, like choice of fertiliser or ways of selecting and managing crops. Notably, the technology-driven interventions are costly and boost the profits of corporations based in the Global North (note that the concepts of Global North and Global South are not meant to refer to strict geographical locations but to the structural and historical inequities between countries and regions of the world in terms of the distribution of power, capital and infrastructure as well as the provision of social services). Does the app bear a responsibility to inform farmers of alternative paths, and of the extent to which different companies may profit from the ways in which agricultural production is managed? Is providing alternatives and context a way to help farmers with their decisions? Or does it generate confusion, thus defeating the very purpose of the app to provide immediate, easy-to-follow instructions? Who owns the app, and does it matter that the underlying code is not available as Open Software for others to scrutinise and re-use?

The questions raised by these cases are difficult ones and do not have easy answers. Yet, they are only a fraction of the questions associated with the technological and social life of data and the ways in which data affect human life. This book aims to provide you with instruments to identify such questions and articulate your own answers to them. It will help you understand the conditions under which data are generated, circulated, traded and used. It is not a technical book about **data science** and artificial intelligence. Rather, it is a book about the interactions between the social and technological aspects of data work.

Before we start, we should say a few words about our backgrounds and motivations for writing this book. We are both scholars working in the broad area of science and technology studies (STS). Anne Beaulieu's background is in science and technology studies and Sabina Leonelli's is the philosophy and history of science. This book project emerges from our experiences in creating Data Wise: Data Science in Society, a minor programme at the University of Groningen; and the training in Data Ethics and Governance for all the post-graduate and professional Data Science degrees at the University of Exeter. The book also builds on our decade-long collaborations with data practitioners working in top research programmes around the world, as well as policy-makers working on **data governance** in a variety of national and international contexts. Through this work, we realised that teaching material concerning data and society at both the undergraduate and postgraduate level was sorely missing. While the academic literature in so-called 'data studies' was blossoming, there was no obvious textbook available to introduce students to this emerging field, and thus complement the technical aspects of data science teaching with an introduction to its social components. This book was conceived to fill this gap and support teaching and learning about data in context.

Part I

DATA IN SOCIETY

Summary

This part frames the critical exploration of the production and uses of data in society. It introduces the notions of '**datafication**', 'data work' and '**data journeys**'. It explains their components and illustrates where they arise across different areas of social life. A number of prominent myths around **Big Data** are discussed and the various characteristics of data are described. We introduce the cycle of knowledge production and show how this cycle helps to better define data and how to understand its roles. The section concludes by showing how data work involves many seemingly small decisions that have ethical implications.

Learning objectives

This part will help you to:

1. understand datafication processes and components;
2. analyse how knowledge, data and technology relate;
3. appreciate that different kinds of knowledge exist across time and spheres;
4. identify how and where knowledge and its trustworthiness arise as issues in everyday life;
5. understand the pervasive role of ethics in data journeys.

1

DATA IN SOCIETY

Summary

Most contemporary societies privilege ways of knowing that are grounded on data. Data have not always been so important, but now play a central role in all kinds of important expertise and decision-making processes. In order to understand and use data, it is useful to develop an awareness of what data are, what they can do and what they should do. This chapter provides a number of starting points to deepen your knowledge about data and learn to better evaluate it. It introduces the concept of datafication as a layered approach, sets out the current context in which data has come to matter and discusses the importance of considering aspects beyond technology to evaluate data and its role. The prevalence of ethical decisions across all aspects of data work is explained.

1.1 Introduction: Who cares about data?

A key development in recent years has been the increasingly prominent role of data in society. Most activities and interactions that any one individual has in contemporary high-tech society produce traces and generate data; whether using email, shopping online or browsing the internet. This is sometimes called the 'datafication' of society – that is, the process through which human activities leave a digital trail. Datafication has two interrelated components: the creation of a trace that is recorded and circulated in the form of data beyond that particular moment and place, and the further use of such a trace as a meaningful element in other processes. Importantly, datafication is not always evident in our everyday life, and many of our activities are datafied whether we wish it or not. For instance, when watching a movie on a streaming service, data are created that document our choice of movie, the time of day at which we watched it and whether or not we watched it all in one go. These data are by-products of our activity: we are generally not watching a movie in order to produce the data, and we may not even be aware of that happening. Yet, these data are generated and used in order to understand and predict our behaviour. The label 'Big Data' has been used to refer to the assemblage and use of vast amounts of data created in a variety of different ways, but this covers only some aspects of datafication, as we will see in the next chapter.

Whereas data were long considered to be a by-product of scientific research, they have now become an output in their own right, in research and in many other spheres. Data, in other words, are now a social phenomenon. This is an important reason why data are often referred to as a singular entity in the popular press and in everyday usage of the term. Data has acquired a reputation as an uncountable, a collective noun for an undefined entity. Everybody recognises the significance of data, but the material and substantive features of data are hard to pin down and understand. When we talk about data as a singular noun, the individual data point no longer matters. What is relevant are the emerging practices in which masses of information underpin decisions, knowledge claims and social perceptions. In this sense, 'data' has come to exist as a fruitful concept in social, political, technological and corporate settings. This is the sense in which we talk about data as 'the new oil', or the 'data deluge', or living in the 'age of data'. This is rather different from the understanding of data as a set of observations or measurements to be used as evidence. When referring to data as a phenomenon, as an area of concern for particular social debates or as an object of study for specific areas of scholarship, we will therefore use the singular in this book.

This is not, however, the only way to understand data. A major reason for scientists to think of data as plural is that there is usually little value in an isolated data point. Data are multiple, collections of objects that are assembled and – crucially – disassembled and reorganised depending on how one wishes to use them. Think of the data that Google holds about users of its email, searching or streaming services: while there may be so many data as to defy human comprehension, each and every one of those data points is important. All these data refer to someone's address, age and music preference, for instance. Each data point might matter in its own right and can be combined with other data points to fuel different types of inferences – from the size of clothes one may be shopping for online, to the locations of events that may be of interest to a Google user. When we think of data as a plural noun, we highlight the way in which data are constituted, assembled and traded, and the judgements and intentions underpinning any one way to cluster data for use. Where data come from, how they are generated and valued, who works with which data and why: these aspects are indispensable to identify and critically evaluate the multiple roles of data in society. For this reason, we use the plural when we want to stress that data are created in specific contexts, and that they have a provenance, quality, quantity and form, and that they are handled and connected in particular ways. In sum, 'data is' refers to data as a phenomenon, 'data are' refers to a collection of data points.

Scientists, journalists, business people, politicians, policy makers and governmental institutions all make use of 'Big Data' and 'data-driven approaches' to understand our society and to shape our daily lives. These data come from somewhere, and the means through which they are collected, circulated and analysed strongly affect how data are ultimately interpreted, and for which ends. The first aim of this book is therefore to *unearth the conditions under which data come to have social value*. This includes all the stages of data handling: from the generation of data to their dissemination through databases and infrastructures, their visualisation through models and interfaces, their combination, through to their analysis and interpretation across multiple settings.

Data also go somewhere, and in fact the more they circulate across a variety of settings, the more value they tend to acquire. Data that get stuck within the walls of one laboratory, one company or one government agency will be used much less, and be interpreted in a narrower way, than data that travel beyond those walls and are scrutinised by more and more diverse people, analysed through various different tools and integrated with data coming from other sources. At the same time, the role of data is not uniform across contexts. There are significant differences in how data come to matter. From sports to healthcare, from business to biology, from social media to education – data and data infrastructures have become a dominant force in all these spheres. Yet, what proves a tremendous opportunity in one domain may well raise significant challenges in another. Hence the second aim of this book is to *provide tools to identify the diverse journeys of data, understand these differences and use that understanding to guide the management and use of data*.

Who cares about data, their provenance and their journeys? We imagine the audience for this book to include not only data scientists and curators (i.e. people whose main responsibilities involve the analysis and stewardship of data), but also anybody who needs to manage data as part of their work, be this in industry, policy, social services or any other profession. It is not just data experts who need to care about data and their social role. Anyone whose job includes collecting, managing and/or interpreting data needs to worry about the implications of their data practices for their business and society at large, and to acquire skills that will empower them to make sensible decisions in that respect. The main audience for this book is what we will call '**data workers**' (see Figure 1.1). Data workers are individuals who may or may not have technical abilities and be directly involved in the development of data analytics, but who are in a position to take decisions concerning what data should be gathered, for which purposes and in which ways. This book will also support the work of those who decide who owns the data and whether it should be shared (and with whom), and those who decide on how to reuse or repurpose data and on whether further analysis may be appropriate and justified.

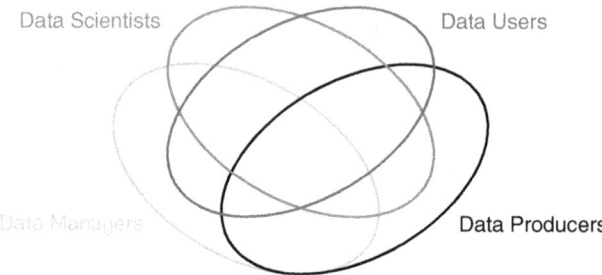

Figure 1.1 Engagement with data: different roles of data workers that often intersect

Source: Figure realised by Michel Durinx, copyright of Anne Beaulieu and Sabina Leonelli.

1.2 Datafication and its components

Many Big Data advocates have discussed the datafication of society as centred on the acquisition and technical handling of data (e.g. Mayer-Schönberger and Cukier, 2013). These data have several features, typically including the increased volume of data at hand, the velocity with which they are produced and analysed, and the variety of data types and sources. These discussions often use the labels of the 3Vs (volume, velocity, variety), or modified versions that add further Vs (venue, vocabulary, vagueness, validity, veracity, etc.). This focus on data and its features misses necessary aspects of datafication as a process. In other words, it is not just about the data, it is also essential to understand how data comes about, what is done with it and why, and to whom this might matter.

Datafication is in a first instance, the turning of objects and processes into data (Mayer-Schönberger and Cukier, 2013). As Van Dijck and colleagues (Van Dijck et al., 2018) show, this 'turning into data' has a number of dimensions. Think of how networked **platforms** render into data many aspects of the world (and our behaviour in it) that had not been formalised in this way before. For example, social networks have become formalised on social media platforms, via the rendering of social ties as digital traces. Our 'friends' on Facebook or the accounts we follow on Twitter are recorded as digital data. Datafication is also the process of rendering activity as quantifiable traces in which patterns can be discovered. For example, the platform LinkedIn makes it possible to create an individual profile, including a photo, and to list one's employment history. LinkedIn keeps track of when users update their profiles and when they change their employment data. The company has identified patterns of activity on users' profiles (such as updating the profile photo) that indicate that an account holder is likely looking for a job and is therefore a good target for job ads. Datafication is also the transformation of interactions into data that can be valued and used for predictive activities. Examples of this are the analysis of public sentiment from tweets and the prediction of electoral outcomes; or the tracking of population movements via localisation of smart-phones, and its use to predict and prevent the spread of disease as in the case of COVID-19. Datafication is finally also the extension of the collection of traces for every interaction, even seemingly trivial ones like the direction and speed of a cursor moving over a webpage or the number of corrections in a draft message before it was posted or published. Some examples of this are often based on surprising **correlations**, such as the relationship between a user's clicking speed and depression.

What makes these processes possible? The extension of automation, with the proliferation of digital technologies, the willing production of massive amounts of data, and the combination and mobilisation of datasets are all important (Rieder and Simon, 2016; Van Dijck et al., 2018). But technological possibilities are not sufficient to explain the scope of datafication. As soon as we consider the various environments and practices involved in making and interpreting data, it becomes clear that datafication is not only about data and related computational techniques. For this reason, we propose a layered model that puts neither data nor technology at its centre. Datafication is a practice that links at least four necessary elements:

1. The community of actors (and related institutions) who engage with the data, for example because they handle the data on an everyday basis. Many forms of engagement with data are possible: some may use it, innovate with it or even oppose it (see Figure 1.2 on the various spheres of datification). This community could be constituted of one or many social groups and have various degrees of cohesion depending on whether or not those involved know each other, have shared values, backgrounds and goals, and work in similar conditions. For instance, a research group working within a small academic field may be highly cohesive, since all its members are likely to have received similar training by the same mentors, and share an interest in – and understanding of – a narrow and well-specified range of topics. By

contrast, users of a fitness app may vary considerably in their values, interests, skillset and background. Given how widely some data tend to travel, and how differently they can be perceived in different parts of society, the community relevant to a particular dataset could be so dispersed and unbounded as to be difficult to identify. Without actors, however, datafication will not be a dynamic practice. When we talk of solutions looking for problems, it is often the case that there is a lack of engagement on the part of actors.

2. The forms of care to which data are subjected, which include specific ways of attributing meaning to data, regulations and laws aimed at preventing data abuse, as well as incentives to value data in particular ways (from the affective, for instance if the data concern a loved one or a cherished project, to the financial, if data are the result of a big investment and/or promise to deliver significant profit). This also includes the care work needed to keep data viable (maintenance, repair, back-ups, etc.).

3. The capacities of those who handle data, whether they are humans (in which case it is a question of different skills, training and experiential baggage) or machines (in which case, we are looking at statistical methods, computational tools and **machine learning algorithms**, as well as hardware such as storage and dissemination systems, **networks**).

4. Data themselves in their many forms (ranging from numbers to images, text, symbols), whose significance and value depend both on their physical characteristics and on the care and meaning bestowed upon them by those who use them.

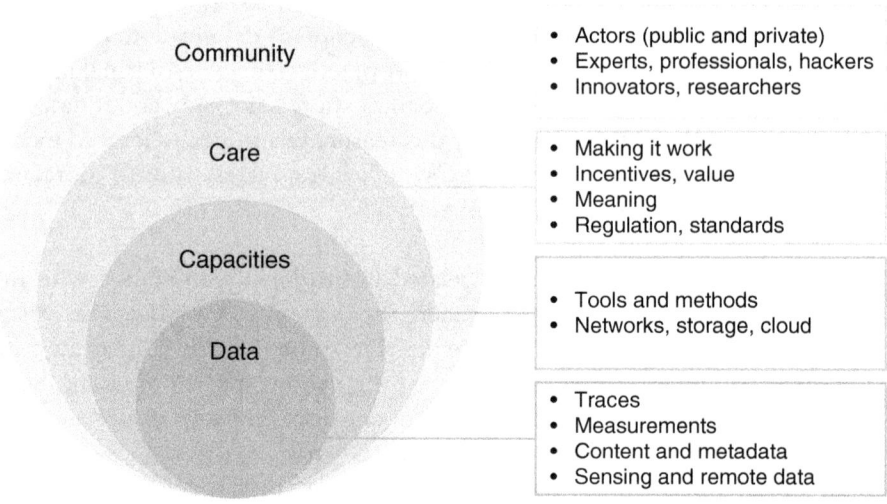

Figure 1.2   A model of the spheres of datafication

Source: Adapted from the ecosystem of Big Data (Letouzé, 2015) through the addition of the care sphere and explanation of the meaning of each sphere.

Datafication links community, care and capacities to data. This link high-lights how data, big or small, never stand on their own. Data only matter if someone cares about them and takes care of them. Care includes all the ways in which we value, regulate, curate and give meaning to data. As we will illustrate in Part II, without users who are interested, who have relevant expertise and who engage meaningfully and creatively, there can be no data in the first place. Because care is so complex, it requires an entire community of data workers.

To illustrate how data is not enough and how datafication requires these four elements, think of recreational running and how that sport has changed in the past decade. Many runners now use a fitness watch that measures and tracks various aspects of their performance such as distance, speed, heart rate and geographical path. These are the data, the measurements whose value relies on a broader landscape of capacities – such as WiFi networks, servers, applications for displaying the data to the runner and digital platforms where the data is shared. In turn, these capacities require care. The manufacturers of the watch need to coordinate their efforts with platform providers so that their services are smoothly integrated. The runner needs to charge their digital tools appropriately, maintain their accounts on the platforms, and check the quality and reliability of data against their own experience. They may also increasingly care about what the data can and cannot say about their running, and train in ways that accommodate the parameters used by the fitness watch (i.e. by training according to heart rate zones rather than perceived effort). In this, the runner may be assisted by a community, made up of a coach who uses the data to evaluate the quality of the training, and of other runners who socialise on the platform because they share data via the same app. This com-munity may be targeted by marketers who focus on particular types of runners based on their data profiles.

It is useful to think about how these components of datafication intersect in the case of other social practices and experiences. To focus only on data and their features is to miss a huge part of the way data come to matter. Weight loss, child rearing, wayfinding, driving, political debate, medical treatment, opinion formation, entertainment, travel: all have been reshaped by datafica-tion. The stronger the ties between the four components of datafication, the more thorough the process of datafication.

Datafication shapes everyday lives in a variety of ways. It shapes self-presentation to others, as evidenced by the attention paid in managing our social media profiles. For example, someone might post information in par-ticular ways that are likely to yield the kind of profile and garner the kind of attention she or he considers desirable. To do this means taking into account what we know about how algorithms and platforms work or how other users react. Datafication and the increased entwinement of various digital settings also mean that different audiences might merge (Beaulieu and Estalella,

2012; Pitcan et al., 2018), and the extent to which profiles and presentation can be managed is also limited. Datafication can also create new vulnerabilities: when undocumented immigrants become more visible because they use digital platforms to find work, they may be more easily exposed to surveillance and detection by immigration authorities (Ticona, 2016).

Indeed, it is important to keep in mind that not all practices undergo datafication in the same way, or to the same extent. As the examples discussed in this chapter show, datafication ranges from the personal (the datafied runner) to the global (Internet of Things (IoT) agriculture). Datafication becomes well established, even entrenched to specific ways of life, only when it is strongly established across data, capacities, care and communities. Later on, in Chapter 3, we will also consider how the qualities associated with data tend to make us value **metrics** and indicators in many areas of knowledge production and policy making.

1.3 Data, ethics and knowledge production

When you use social media, you trust the company that maintains the specific app you are using with all sorts of personal information. You may be sharing contact details of friends and family, your favourite places to hang out or the events you attend. Is it okay to use those data to inform traffic control so that you can more easily go to those places? And what about using data about your friends to work out whether you are an outgoing person or not, or to predict which political party you are likely to vote for? Besides this kind of data that we explicitly share, much more data is gathered about us, from our patterns of logging on to a platform, to the frequency of our likes or the length of our replies. Using your data to acquire insights about your behaviour and preferences can lead to what seem to be unambiguously good social outcomes – such as better traffic flow, easy access to services and increased safety. And yet, even in that case, the release of location data from smartphone users can have unpleasant, unexpected side effects. For instance, it can highlight places and times in which vulnerable people find themselves alone and isolated, thus facilitating stalking and attacks. Other outcomes, such as more accurate polling before an election, are more ambiguous in their social effects. As exemplified by the scandal surrounding how Cambridge Analytica used Facebook data to launch aggressive political advertising, better predictions of voting behaviour can help politicians to tailor their policies. It can also interfere with public discourse in ways that may be perceived as dishonest and ill-intentioned. These are cases of concern to **data ethics**, which informs the evaluation of what constitutes right and wrong actions in relation to data handling. Data ethics is complex and ever-present. It is part of all design and selection decisions and practices around data, and it typically involves choosing between different options in the absence of obvious solutions or even without knowing what the results of such choices will be. In fact, data workers have to make choices even in situations where all options are problematic, and/or there is no clarity over what constitutes the 'right' choice. Crucially, data ethics is not just a conversation happening on the sidelines of data work, but rather underpins many actions taken in the course of such work (even when the action is to do nothing) – a fundamental characteristic to which we return in Chapter 9.

A key concern in data ethics is the open-endedness of data use. It is not always possible to say ahead of time exactly how data will be used and with what effects. Technologies for the production, dissemination and analysis of data keep evolving at great speed. This is partly due to the scale of investments by governments and corporations and to the lack of regulation of these activities. These technologies are often used in unpredictable ways. The IoT is increasingly dominating human existence, and data-driven systems are entering spheres as diverse as policing, immigration, healthcare and energy consumption. It is ever more difficult to know which data are held, by whom and where. All these elements add to the difficulty of knowing how data could be used in the future, and what the social, economic and political implications could be. How we deal with this complexity is an ethical question. While it is, by definition, impossible to deal with all implications of data present and future, it is possible to reflect on the best ways to address dilemmas, and on which values should prevail when faced with these. This book aims to show the complexity of such processes and to help deal with effects in more responsible ways.

For example, in 2008, European Union (EU) policy decreed that every household in Europe should be equipped with a smart meter that can transmit data about energy use and production. Armies of engineers subsequently installed new devices in millions of homes. And yet this massive change was not accompanied by a discussion of the different roles that households, which now have more data about their energy use, could play in new energy regimes. Nor has there been a public debate over the trade-offs between the related loss of privacy due to the smart meter, and a better-managed power transmission grid. Fostering such debates does not require full knowledge of the possible consequences of datafication, nor does it involve reaching consensus on ways forward. What it does require is providing a space to consider possible implications from a variety of viewpoints, and ensuring that potential concerns are explicitly acknowledged and taken into account in the development of technology-focused social interventions.

A related concern is the opacity of the technical work surrounding data processing, analysis and interpretation. As we will see in this book, many decisions with significant ethical implications are taken in the course of developing systems for data analysis. Simple decisions, such as whether to accept a given data format or source, or how to label particular datasets have implications for who and what counts and who or what are made invisible. The extent to which data analysis depends on statistical, mathematical and computational expertise makes such work daunting for anybody who is not trained in these fields. In turn, this creates the fear that technicians may implement ethically dubious decisions without any oversight or consultation. And yet, the idea that a statistician or a computer scientist would be able to fully comprehend data systems is also misleading: many different forms of specialised work are required to process and analyse data. This means that even very technically savvy data workers may not be able to understand the data system as a whole, and much less to predict its implications.

Does this mean that we cannot do anything about ethical issues? We think that being aware of this complexity actually encourages us to have regular and wide-ranging consultations on the possible implications of technical changes to data systems. Consider for instance the growing tendency to encourage users to make use of a Facebook, Twitter or Google account to access other digital platforms. Think of using a Facebook account to sign into a library account. While accepting this service may seem harmless and even convenient, it has big implications for data flows. By using your Facebook profile to log onto a different service, you are linking two sets of data. Different databases operated by different platforms become linked. This enhances the value of data held by the corporate owners of the platforms – in this case, social networks and book and media use can be correlated. In addition, reusing profiles across different platforms makes you more vulnerable to identity theft or other privacy breaches. This practice of using profiles to log onto other platforms also makes it more difficult *not* to use particular platforms. If the expectation is that one will use an existing profile, accessing the platform in an alternative way increases the cost (in time, in attention, in amount of work required) of *not* participating in a given platform.

To realise how these seemingly mundane practices have an ethical dimension is not based on in-depth technical knowledge. Seeing the ethical dimension requires a basic understanding of the system, coupled with the opportunity and time to think about its implications. In a world where technological development (or innovation) is considered both a good in itself and a competitive advantage, such opportunities and time are seldom created. This increases the perception of technological choices as impenetrable. So how can we ensure that there is attention to ethical aspects of data? The field of data ethics focuses on this question. The most important goal of data ethics is to promote responsible and sustainable data work in ways that may contribute to human flourishing

(Floridi, 2014; Floridi et al., 2018). And yet in the context of datafication, where intricate flows of data and multiple sets of algorithms are the rule rather than the exception, it can be very difficult to establish who is responsible for ensuring that data are used ethically. This can mean not knowing who should take the blame when things go wrong, or who should be responsible for fixing errors or solving problems when they arise. Incorrect data can be difficult to remove or correct once they have moved beyond the context where they originated. Imagine an error in a school record that is shared with an employer, and then travels to a person's file at an insurance company. It can be very difficult to trace where the error occurred and nearly impossible to find out where the incorrect data ended up. This is one of the reasons why governments are currently working very hard to articulate new laws that clarify responsibilities in relation to data. In this example, personal data is at the forefront, but as we will see in Chapter 10, attributing responsibility is even more difficult in cases where data about groups get built into data infrastructures or where machine learning tools are trained on historical datasets.

In nearly every aspect of data work, ethical decisions will be required, and there will be no readymade answer as to which course of action is the best. This book will treat data ethics as an integral part of data management, and point you towards tools, principles and guidelines that can help to identify, address and resolve ethical concerns as they arise in data handling and use. This will be the main focus of Part III. Furthermore, across the chapters in this book, we create opportunities to stop and think about the small moments of design and decision-making and about the effects of large-scale implementation. By discussing ethical consideration across many different situations in relation to data, we hope to enhance awareness that **ethics** is not a one-off or separate kind of concern. Data ethics contributes substantively to the effectiveness and positive impact of data solutions.

1.4 Conclusion: The Impact of Datafication

What is happening to knowledge in contemporary society is not simply that we have more or 'bigger' data: the whole system of knowledge production is changing. Datafication involves not only data, but also community, care and capacities – all of which rely on material conditions, values, preferences and norms for acceptable behaviour. When knowledge practices integrate data, they also align across all components. For data to matter and become evidence for a specific claim or course of action, all four have to be in place.

Data-intensive practices connect to contemporary ideas about what is good knowledge. Put more concretely, we are living in a world where data is part of what we feel we need to know in order to parent, police, govern, be healthy or put together a good soccer team. These changes in knowledge are related to issues of trust in knowledge and truth. As data becomes more central to how we produce knowledge, data also becomes more central to how we evaluate knowledge. This holds for everyday knowledge queries (we google to find out), for researching life-changing decisions about health or real estate (we look for data to inform our decisions; we compare our data to averages/profiles/percentiles) and for participation in public life (we discriminate between real and fake news). In all these activities, we use our data skills and insights – what we know about how data is generated, what is good data, how to analyse data responsibly and how data might be tampered with. All these ways of evaluating data are far from self-evident, if you look back in time just a couple of decades. In the chapters that follow, we will explore different aspects of data work and provide a set of concepts and tools to further think about, analyse and evaluate data.

ADDITIONAL READING

Ebeling, M. (2016). *Healthcare and Big Data: Digital Spectres and Phantom Objects*. London: Palgrave Macmillan.

Floridi, L. (2014). *The Fourth Revolution: How the Infosphere is Reshaping Human Reality*. Oxford: Oxford University Press.

Kitchin, R. (2021). *The Data Revolution: Big Data, Open Data, Data Infrastructures and Their Consequences*. London: Sage.

Van Dijck, J., Poell, T. and De Waal, M. (2018). *The Platform Society: Public Values in a Connective World*. Oxford: Oxford University Press.

Part II

DATA CREATION

Summary

This part focuses on the creation of data. It discusses our expectations of what data are and what data can do. We start by discussing the promise of Big Data and the historical development of data in Chapter 2. Chapter 3 reviews the characteristics of data, and the importance of context of creation and of data journeys in shaping the meaning and characteristics of data. By discussing concrete examples of data creation and data journeys, we show how taking these elements into account puts us in a better position to evaluate the suitability and reliability of data. In Chapter 4, we turn to different ways of conceptualising data, and contrast the representational and relational view of data. A fixed, representational view of data positions data as a foundation on which to build knowledge and emphasises the need to remove bias and noise as the most important data work. A relational view positions data as changeable and contextual, and emphasises that many kinds of data work are needed across all steps of knowledge production. Finally, we consider the changing role of data, as it becomes more central to how we evaluate knowledge and to the broader knowledge production cycle.

Learning objectives

This part will help you to:

1. understand the historical roots of data science and Big Data;
2. understand what data are and how they relate to knowledge production;
3. critically evaluate hyperbolic claims on the power of Big Data;
4. identify the various technical, epistemological, social, legal, institutional and economic dimensions of data journeys and of how they are entwined;
5. evaluate data according to their provenance, their merits and disadvantages, and critically assess their quality and limitations.

Data Story 1: Big Data on Consumer Habits

The introduction of credit card payments and loyalty cards for customers of supermarkets has created a vast amount of digital information on customers' preferences and spending habits. These technologies are typically advertised as means to facilitate payments and obtain discounts, so customers do not necessarily think of these technologies as tools to produce data. Indeed, the primary function of credit cards has long been to facilitate payments, and the function of loyalty cards is, at least on the surface and as their name suggests, to ensure loyalty of customers to a store or brand, usually by providing special offers. At the same time, such technologies have created vast amounts of data about customers, which provide a wealth of insights to supermarket owners and retail companies. The data can be used to identify purchasing patterns that can indicate which products are most popular, at what time of the day or season, and among which types of customers. These analyses can support marketing strategies as well as helping to manage the supply, distribution and shelving of items.

- *What kind of data are these?* What makes the digital traces left by supermarket transactions into 'data'? What characteristics do such data have? What are these data about?
- Can you think of any disadvantages of using these data to inform product supply? Is the use of these data always beneficial? Who benefits?
- *Does the scale of data collection make a difference, and how?* For instance, does it matter whether we are considering data collected by a large supermarket chain with hundreds of sites around several countries, or whether we are considering data from three local shops whose customers may know each other?

The combination of data from supermarket transactions with other types of data can provide even better insights. Such combinations become possible when customers sign up to a given loyalty scheme, and typically provide their address, date of birth and telephone number. Further combinations may be the result of doing searches of customers' names and finding out their medical history, personal preferences and lifestyle habits from social media. The combination of these data sources gives rise to an enormous data pool, often called 'Big Data'. From such a data pool, data analysts can extract predictions about what a customer is likely to buy in the future, how their social lives will evolve or where they are likely to go on holiday.

- *How do Big customer Data work?* When does a given dataset deserve to be called 'Big Data'? Why are these data valuable, in which ways and for whom? When different types of data are combined, do they increase the accuracy and reliability of the knowledge being produced?

An improved understanding of the customer base enables many companies to tailor their services closely to the desires and needs of their clients. It also increases the opportunities to manipulate customers' behaviours via targeted advertising or special offers geared to facilitate addictions to specific types of products.

- *Does it matter whether customers are aware of who is using their data, and how?* Why? Are data derived from consumer services or social media always reliable? Can you imagine cases where the knowledge extracted from such data is not trustworthy? To what extent can repeated suggestions and nudges from companies shape our behaviour? Is it possible to ignore this targeting, and if so, to what extent?

Data Story 2: Remote Sensing for Conservation Research

Remote sensing technologies such as drones are widely viewed as the new frontier of data collection, especially when it comes to environmental and biological research and monitoring. Among many other things, they help to map the spread of crop diseases around the world, the extent of deforestation in the Amazon and the degree of damage to coral reefs. In this second Data Story, we look at the use of 'unmanned' aerial vehicles (UAVs) in conservation projects to detect wildlife and monitor the behaviours of protected species. In particular, we consider how UAVs have been used to produce data documenting the location of chimpanzee nests in Tanzania.

As with other species of great apes, chimpanzees' survival is heavily threatened by environmental changes, deforestation, disease and poaching. To help protect this species, conservationists agree on the need for accurate data on their distribution and density, which need to be gathered at regular intervals. These data can help to identify areas where habitat encroachment, poaching or disease are leading to population loss, thus paving the way for targeted interventions. Given the sheer scale of chimpanzee distribution across western Tanzania (over 20,000 km^2), there is urgent need for cost-effective methods of data collection and analysis that can reliably and frequently track chimpanzee numbers and movements across broad spatial scales.

Recently, drones have been used to identify nests from the air. Drones fitted with cameras gather images by flying over large areas. To identify nests, the photographs and videos gathered by the cameras can be analysed automatically using machine learning. This is more effective than using aircraft or satellite data, since those tools do not have sufficient resolution to be able to detect smaller animals

(Continued)

17

and their traces. Using drones seems to require much less effort and resources than surveys conducted by technicians on the ground. A key complication is that it is actually difficult to observe chimpanzees themselves through these data. Indirect **evidence** of their presence (e.g. dung, calls or nests) is typically the main parameter used to estimate population size. Much indirect evidence is gathered on the ground by specialised technicians, often using ground surveys. For instance, a 'line transect survey' is done by counting all individuals in one specific strip of territory and using that sample to estimate population size over the whole area. This way of acquiring evidence is arguably much more labour- and time-intensive than the use of drones.

- *What kinds of data are acquired via observational techniques like ground surveys?* Objects used as evidence are not always numbers resulting from measurements: they may include pictures of footprints, samples of droppings and handwritten notes about sounds heard in the forest. Are these the same kind of data as the images produced by drones? Are the data obtained from drones comparable with data acquired through observations from the ground, and how?
- *How can we evaluate the quality and reliability of data in this case?* Is quality evaluation possible in the absence of multiple sources of data? For example, would we be able to evaluate the quality and reliability of drone image detection without triangulating with data acquired on the ground? Are drone data 'good enough' to warrant stopping ground surveys altogether, thus generating enormous savings for conservation efforts? Or should the drone census be combined with a survey on the ground to confirm the results (which may increase the accuracy and reliability of results, but will also increase the costs)? How could these data be complemented by tracker data to understand habitat use and range?

Let us now consider whether the use of drones actually does cost less effort and resources than the use of ground surveys. Drones can be operated by a small team that needs to travel to the various national parks being investigated. To use the drones, at least one member of the team must hold a drone pilot licence (which is expensive to obtain) and be knowledgeable about demarcated flying areas, flight height and privacy laws. Within each location, the team members need to adjust and calibrate the cameras, set them up on the drones and wait for the right weather conditions to fly the drones according to a predetermined grid pattern. Depending on weather changes and visibility conditions on the day, the data may be more or less consistent, and the instruments will need to be regularly recalibrated and cleaned up by the data scientists who are in charge of integrating results collected across different sites. Different teams may also use different parameters to calibrate the cameras or set up camp, which again will determine some differences across datasets that will need to be evaluated manually.

- *Which data are most expensive to produce and analyse?* When taking the data work required to make sense of drone data into consideration, does it still make sense to view them as less resource-intensive than survey data? In which ways do these differences matter to the broader effort of designing, implementing and financing a study?

Now, consider that when conservation officers do a ground survey, they walk through areas that they are very familiar with. They know where to look and detect all sorts of signals from the environment, some of which may come to acquire significance later – for example, in case photographs of local trees reveal symptoms of a newly emerging plant disease. They also talk to visitors and inhabitants of the park. Conservation officers can thus pick up signs of poaching and raise alerts, spot new species moving into the area or identify changes in behaviours of local animals. By contrast, the drone team, while technically savvy, may not have the same degree of familiarity with the area since they are only occasionally present and tend to cover much broader territory in much less time. The imaging data that they collect can reveal all sorts of unexpected things, but they come in a highly standardised format and there is little chance to deepen observations or investigate unexpected findings on the spot. At the same time, the use of drones makes data more spatially explicit and has a higher spatial resolution (geolocation of less than 1 m) than human observations (about 15 m accuracy). Aerial surveys can also be done more often, since line transect surveys on the ground are very time consuming.

- *How do choices about data acquisition affect the type of research – and knowledge – subsequently produced?* Given these different ways of working, how do you expect the resulting insights about chimpanzee populations to differ? Are these differences a problem? How might the results of the two teams be evaluated by other conservationists, researchers or policy makers? How much does precision matter? What are the advantages of a cheaper and faster identification of nests?
- *How do choices about data acquisition affect what is valued as relevant expertise in a project?* Does reliance on drone data favour the deployment of technical personnel (no matter where they came from) over people with local knowledge and familiarity with the areas in question? Who is best qualified to be doing the identification of animals and estimation of population size and state? Does it matter if knowledge is not produced by people who have ties to the areas, and how?

(Continued)

The image detection system can identify nests, though researchers feel that this could be improved even more with better image resolution. When nests are in the middle of the tree crowns, they are more difficult to identify using drones and many nests are missed. To increase detection, researchers want to improve image resolution by using lower altitude drones with more sophisticated multispectral cameras. This use of the drones is likely to have a higher environmental impact, as drones would be visible to animals and their passage may affect the local ecosystem. Also, were these cameras to be flown over inhabited areas, they would capture minute details of the everyday life of humans living there.

- *What are the broader implications of choices made when creating data? What could be the effects of improving detection in this way? What could be the effect on the data gathered? What could be the effect on the chimpanzees? Who should decide whether the increased resolution and nest detection are worth possible negative effects? Could this be taken into account in the design of the tools?*

Data Story based on: Bonnin, N., van Andel, A. C., Kerby, J. T., Piel, A. K., Pintea, L. and Wich, S. A. (2018). Assessment of chimpanzee nest detectability in drone-acquired images. *Drones*, 2(2): 17. https://doi.org/10.3390/drones2020017

2

BIG DATA IN CONTEXT

---------- Overview of chapter ----------

Summary

In this chapter, we focus on how and why Big Data has become so prominent. We review the high expectations associated with Big Data and examine 'Big Data mythology', the often overly optimistic views of what data can and will be able to do. We show how Big Data is part of a longer history, in spite of being hailed as a radical innovation and we highlight the limitations of Big Data. The long history of data across different social circumstances and periods makes clear that what we think of as data, and what we think it is good for, has changed radically over time.

2.1 Introduction: The rise of Big Data

The datafication of society is characterised by three main features. First, we see that the creation of data is becoming important and increasingly valued. Second, by using, combining and visualising data in everyday life, data become even more central. And third, we take more and more decisions about current and future actions based on data. If we formulate these three features more conceptually, we can say that (1) data are viewed as valuable commodities that have the power to transform society, and they are no longer viewed as by-products of administrative and research processes; (2) efforts to mobilise, integrate and visualise data are viewed as central contributions to social life, since the more data are pooled together, the higher the chance that they will acquire new significance and meaning; and (3) the consultation of data resources, typically mediated by complex infrastructures and databases, is regarded as the first step in any process of inquiry and plays a key heuristic role in determining future directions for social action.

Together, these features show how the role of data can become increasingly important. The emphasis on the key role of data as starting point for inquiry is rooted in the wish to capitalise on the 'data deluge' generated by new technologies through the datafication of human activities. The resulting 'Big Data' is a resource for research, with ever more sophisticated computational tools being developed to extract knowledge from such data. The Data Story at the beginning of Part II discussed the case of consumer data garnered from credit card payments and loyalty schemes. Such data can improve current understandings of the nutritional status and needs of a particular population, particularly when combined with data coming from public health and social services, such as blood test results and hospital intakes linked to obesity. Other examples are the use of various different types of data acquired from cancer patients, including genomic sequences, physiological measurements and individual responses to treatment, to improve diagnosis and treatment. Or think of the integration of data on traffic flow, environmental and geographical conditions, and human behaviour to produce safety measures for driverless vehicles. By integrating these data, better approaches can be developed that make it possible to promptly analyse a situation and generate an appropriate response – a child suddenly darting into the street on a very cold day and the driverless car swerving enough to avoid the child while also minimising the risk of skidding on ice and damaging other vehicles. In each of these cases, the availability of diverse data and related analytic tools is creating novel opportunities for research and for the development of new forms of inquiry, which are widely perceived as having a transformative effect on society as a whole. In this chapter we will present some of the key characteristics attributed to Big Data, and then we will critique some of those ideas by pointing to the practical obstacles, conceptual problems and social implications involved in the dissemination, aggregation and use of large datasets.

There are multiple ways to define Big Data (Kitchin and McArdle, 2016). Perhaps the most straightforward characterisation is as *large* datasets that are produced in a *digital* form and can be analysed through *computational* tools.

The two features most commonly associated with Big Data are volume and velocity. *Volume* refers to the size of the files used to archive and spread data. *Velocity* refers to the pressing speed with which data is generated and processed. As an increasing amount of data is produced and processed, it seems that we need automated analysis.

It is precisely those two features, volume and velocity, that are however the most disputed features of Big Data. What may be perceived as 'large volume' or 'high velocity' depends on the context. What is a large dataset in some fields is perfectly normal in others. Furthermore, technologies used to generate, store, disseminate and visualise the data change rapidly. This is exemplified by the high-throughput production, storage and dissemination of genomic sequencing and gene expression data, where both data volume and velocity have dramatically increased within the past two decades. Similarly, current understandings of Big Data as 'anything that cannot be easily captured in an Excel spreadsheet' are bound to shift rapidly as new analytic software becomes established, and the very idea of using spreadsheets to handle data becomes a thing of the past. Moreover, there are many other ways to qualify data. A focus on data size and speed does not take account of the diversity of data types used. This may include data that are not generated in digital formats or whose format is not computationally tractable. This shows the importance of data provenance; that is, the conditions under which data were generated and disseminated. By emphasising velocity and volume, we risk ignoring other important elements that shape the interpretation of data, such as the circumstances of data use, including specific queries, values, skills and research situations.

An alternative is to define Big Data not by reference to its physical attributes, but rather by virtue of what can and cannot be done with it. In this view, Big Data is a heterogeneous ensemble of data collected from a variety of different sources, typically (but not always) in digital formats suitable for algorithmic processing, in order to generate new knowledge. For example, Boyd and Crawford (2012) identify Big Data with 'the capacity to search, aggregate and cross-reference large datasets', while O'Malley and Soyer (2012) focus on the ability to interrogate and interrelate diverse types of data, with the aim to be able to consult them as a single body of evidence. The examples of transformative 'Big Data research' given above are all easily fitted into this view: it is not the mere fact that lots of data are available that makes a different in those cases, but rather the fact that lots of data can be mobilised from a wide variety of sources (social media, environmental surveys, weather measurements, consumer behaviour).

This account makes sense of other characteristic 'V-words' that have been associated with Big Data. These other aspects emphasise functional rather than physical characteristics, and include:

- *Variety* in the formats and purposes of data. Data include objects as different as samples of animal tissue, free-text observations, humidity measurements, Global Positioning System (GPS) coordinates and the results of blood tests.
- *Veracity*, understood as the extent to which the quality and reliability of Big Data can be guaranteed. Data with high volume, velocity and variety are at significant risk of containing inaccuracies, errors and unaccounted-for bias. In the absence of appropriate validation and quality checks, this could result in a misleading or outright incorrect evidence base for knowledge claims (Floridi and Illari, 2014; Cai and Zhu, 2015; Leonelli, 2017).
- *Validity*, which indicates the selection of appropriate data with respect to the intended use. The choice of a specific dataset as evidence base requires adequate and explicit justification, including recourse to relevant background knowledge to ground the identification of what counts as data in that context (Bogen, 2010; Mayernik, 2019).
- *Volatility* refers to the extent to which data remain available, accessible and re-interpretable despite changes in archival technologies. This is significant given the tendency of formats and tools used to generate and analyse data to become obsolete, and the efforts required to update data infrastructures to guarantee data access in the long term (Edwards, 2010; Lagoze, 2014; Borgman, 2015; Sterner and Franz, 2017).
- *Value* points to how – and to what extent – data come to matter to different sections of society. This is not just in a financial or economic sense. Rather, the attribution of 'value' to data encompasses any way in which data could be perceived as significant, whether this is scientific, financial, ethical, reputational or even affective forms of value (Leonelli, 2016a; D'Ignazio and Klein, 2020). Value depends as much on the intended use of the data as on historical, social and geographical circumstances. It is important to note that it is not only individuals or groups who attribute value to data. Institutions, such as the companies and research organisations involved in governing and funding data-intensive science, also have ways of valuing data, which may not always overlap with the priorities of data workers (Tempini, 2017).

This list of features, though not exhaustive, highlights how Big Data is not simply 'a lot of data'. The epistemic power of Big Data lies in its capacity to bridge between different research communities, methodological approaches and theoretical frameworks that are difficult to link due to conceptual fragmentation, social barriers and technical difficulties (Leonelli, 2019). And indeed, appeals to Big Data often emerge from situations of inquiry that are at once technically, conceptually and socially challenging, and where existing methods and resources have proved insufficient or inadequate. Examples range from the attempt to understand biodiversity via integration of highly heterogeneous observations from remote locations around the globe (Sterner and Franz, 2017) to the mass aggregation of social data to track consumer demand for specific products or to produce national measures of economic growth.

2.2 The Big Data mythology: Data transforms society

The emergence of Big Data affects many sectors of society, including scientific research. Promises to enable new and more efficient ways to plan, conduct, institutionalise, disseminate and assess research form a set of expectations that have been embraced by many government agencies, companies and research organisations in the first two decades of the 21st century. They continue to inform the development of analytics and technologies underpinning the use of Big Data. We label these expectations the **Big Data mythology** to emphasise its mythical status and stress the difference between such utopian promise and what Big Data (and related analytics) can actually deliver for society.

The ability to link and cross-reference datasets coming from different sources is expected to increase the accuracy and predictive power of scientific findings, and help researchers – whether they work in universities, industry or policy institutions – to identify future directions of inquiry. The availability of data provides an incentive to build automated procedures and tools to store, organise and analyse the data, in the name of improving the reliability and transparency of knowledge creation. It is widely believed that Big Data is ushering in a whole new way of doing research, which is heavily grounded in data analysis and less dependent on pre-existing theories. This belief is reflected in the renewed attention to data strategies as key component of management for industry. It is also tangible in novel sources of funding and publication venues (such as 'data journals') within academia.

Big Data is often presented as *comprehensive*. This is the claim that the accumulation of large datasets enables researchers to ground their analysis on several different aspects of the same phenomenon, documented by different people at different times. According to Mayer-Schönberger and Cukier (2013), it can become so big as to encompass *all* the available data on a phenomenon of interest. As a consequence, Big Data can provide an all-encompassing perspective on the characteristics of that phenomenon, without needing to focus on specific details. This is a big promise, and perhaps an understandable one, given the speed and extent of datafication in many areas of life. Yet, it is an extreme promise to hold up to say that 'we have all the data'; that everyone or all of reality is captured, as invoked by Twitter's slogan, 'it's what's happening'.

Big Data is also argued to push researchers to embrace the complex and multifaceted nature of the real world, rather than pursuing exactitude and accuracy in measurement obtained under controlled conditions. Indeed, it is impossible to assemble Big Data in ways that are guaranteed to be accurate and homogeneous. Rather, the Big Data mythology encourages analysts to resign themselves to the fact that 'Big Data is messy, varies in quality, and is distributed across countless servers around the world' and welcome the advantages of this lack of exactitude: 'With Big Data, we'll often be satisfied with a sense of general direction rather than knowing a phenomenon down to the inch, the penny, the atom' (Mayer-Schönberger and Cukier, 2013, p.13).

This idea of messiness relates closely to the third key innovation brought about by Big Data, which Mayer-Schönberger and Cukier call the 'triumph of *correlations*'. Correlations can be defined as the statistical relationship between two data values. They are notoriously useful as heuristic devices within the sciences and beyond. Spotting that when one of the data values changes, the other is likely to change too, is the starting point for many discoveries. It is also used to analyse economic activity and in many attempts to understand human behaviour: for any big change in the market, the political situation or the environment (such as an economic crisis, a change of government or an earthquake) changes in citizens' work and spending patterns can be scrutinised to see whether there may be a link. Market research as a whole may be viewed as an exercise in the identification of correlations between a user profile and their preferences for specific products. However, researchers have typically mistrusted correlations as a source of reliable knowledge in and of themselves. This is chiefly because they may be spurious and due to chance – in other words, they may result from serendipity rather than specific mechanisms, or they may be due to factors other than the variables under consideration. For instance, a Netflix user may be using the opening music of a documentary series every night to get his son to sleep, but may never have watched the series and may in fact hate documentaries. In such a case, it would be wrong to interpret Netflix data on his viewing history as being correlated to his viewing preferences.

According to the Big Data mythology, Big Data can override those worries about spurious correlations. In a Big Data world, it is argued, it simply does not matter whether any single correlation is reliable: what matters is the correlations spotted on a very large dataset can help to predict future behaviour with reasonable accuracy. In the example of the Netflix viewer, asking whether he actually liked documentaries becomes irrelevant: what matters is that data analytics can reliably predict that he will be streaming the intro to the documentary again tomorrow. On a much broader scale, Mayer-Schönberger and Cukier give the example of Amazon.com, whose astonishing expansion over the past few years is at least partly due to its clever use of statistical correlations among the myriad of data provided by its consumer base in order to spot users' preferences and successfully suggest new items for consumption (Mayer-Schönberger and Cukier, 2013). In cases such as this, correlations provide powerful predictive knowledge that was not available before, and that can inform society without appearing to be complemented by a causal understanding of *why* a specific effect is predicted. Causal understanding is viewed as simply irrelevant to the useful knowledge yielded from Big Data. Hence, Big Data encourages a growing respect for correlation and prediction, which comes to be appreciated as a more informative and plausible form of knowledge than the more definite, but also more elusive, causal explanation. In Mayer-Schönberger and Cukier's words: 'the correlations may not tell us precisely *why* something is happening, but they alert us *that* it is happening. And in many situations, this is good enough' (Mayer-Schönberger and Cukier, 2013, p.14).

This Big Data mythology is associated with a specific reading of the significance of technology in shaping social life. It is argued that in the past we lived in an analogue world, where computation was extremely limited and mechanical in nature. Now the increased use of information and communication technologies has given rise to a digitisation of experience. In this world, experience that is computable and machine-readable is in turn valued for the possibility of detecting patterns, developing profiles and further predicting behaviour. These practices shape a particular version of human experience – as the sum of our digital traces – as the basis for knowledge. The same holds for expertise. We have seen a shift from expertise as something embodied in a human expert, and developed over time through the active combination and application of knowledge and practice (Daston and Galison, 2007). Increasingly since the 1970s, expertise is embedded in 'expert systems'. These are often considered to be the first forms of artificial intelligence and were developed in the medical field to assist in diagnosis but also in areas as diverse as speech recognition and crisis management. These systems are usually based on formalising the reasoning of human experts as a set of rules. They promise to fully automate processes and eliminate the need for human intervention. In parallel, a culture of metrics and auditing has spread through institutions, industry and governments around the world, in which data is the main tool used to evaluate outcomes and processes. As a result of these changes, the Big Data mythology highlights how automated use of digital data has become central to *how* we know – with medicine, business, policy, education and environmental concerns all focused on obtaining and analysing data, often for predictive purposes. The availability of digital data is viewed as the best kind of proof or as the best basis for taking action (evidence-based policy). We return to the role of data in decision making in Chapter 10.

27

2.3 A historical perspective: Society transforms data

Some of the claims associated with the Big Data mythology described above look perplexing when evaluated from a historical viewpoint. For one thing, reliance on large datasets is not a novel development, just as data are not necessarily digital objects but include texts (observations about the characteristics of specific territories), drawings (reproductions of the morphology of a newly discovered species of plant) and even specimens (fossils). Data have long been the foundation of empirical inquiry, with long-standing efforts to collect and organise large volumes of data in domains such as astronomy, meteorology and natural history (Aronova et al., 2010; Daston, 2017; Porter and de Chadarevian, 2018). Similarly, biomedical research – and particularly subfields such as epidemiology, pharmacology and public health – has an extensive tradition of tackling data of high volume, velocity, variety and volatility, and whose validity, veracity and value are regularly negotiated and contested by patients, governments, funders, pharmaceutical companies, insurance companies and public institutions (Bauer, 2008). The world of research is no stranger to the accumulation of data, and to the quest for ingenious technologies – such as archives, punch cards and statistical techniques – that would facilitate the management and analysis of all that material. In this section, we briefly review key points in the history of research data, to exemplify the ways in which, just as data changed society, social change shaped data.

In the Western world of the 17th and 18th centuries, research data were gathered by visionary individuals, such as gentried naturalists like Charles Darwin and court astronomers like Ticho Brahe. These individuals were backed by wealthy patrons and supported by an extensive network of data collectors, who would roam the globe in search of new biological specimens and gather astronomical and meteorological observations from a variety of different locations. This 'age of discovery' was marked by an extractive approach to colonial expansion: rich European countries were focused on identifying and bringing back resources from the colonies that would extend and consolidate their knowledge and power. The large quantities of data thus accumulated were systematised and analysed through models (think about Kepler's laws describing how planets move around the sun, derived from consideration of the astronomical observations collected by Tycho Brahe in the 16th century) or classification systems (such as Linnaeus' taxonomy of different forms of life, which grew out of the study of specimens collected by explorers and merchants in the 18th century and underpins how we distinguish between species to this day). These approaches to ordering data were valuable because they made data more usable and combined principles of organisation with data, so that these systems had both simplicity and explanatory power.

In the 19th century, data shifted from the work of brilliant individuals to a more institutional approach, with national agencies such natural history museums, boards of trade, the census and weather services emerging as a constitutive part of the administrative apparatus of national governments. This means that data were increasingly recognised as social commodities with scientific as well as financial and political value. Data became something to be invested in, regulated and managed – and was clearly marked as a tool for governance and trade. Again, this accompanied a shift of colonial rule: this time it was from extraction to control, with dominating powers increasingly interested in how to manage large populations in the wake of increasingly intense revolts at the turn of the century, and epidemics – and related food shortages – proving increasingly disruptive to urban life and global trade.

Through collaboration between national information infrastructures, more sharing took place, leading to a new informational globalism (Hewson, 1999; Edwards, 2010). With the rise of nation states and the increasing demands of international trade, these initiatives aimed to measure both nature and society in a more systematic, depersonalised manner, and were fostered by an ever-expanding group of data aficionados including researchers as well as administrators, merchants and politicians.

As data became a growing concern, sophisticated techniques of quantification were developed. Statistics became a separate discipline. As more complex techniques and more experts became available, more complex types of data gathering were developed, such as the census (Von Oertzen, 2018). International entities such as the League of Nations and the International Monetary Fund had clear aspirations to globalise data collection and analysis for a variety of purposes and across all scientific domains: from drug testing, with the creation of the Permanent Commission on Biological Standardisation to monitor chemical tests and biological assays from 1924, to economic assessments through comprehensive data collection on employment, unemployment, wages and migration by the Economic and Financial Section of the newly instituted International Statistical Commission. Population-level thinking gripped the life sciences through the widespread adoption of the Mendelian theory of inheritance, which was fruitfully combined with attention to new types of data – and specimen – collections focused on genetic mutants of the same model species.

Statistics became the main source of information for emerging insurance practices and public health monitoring systems (Porter, 1995; Desrosières, 2010). While statistics now seems to us an obvious way to think about health, poverty or employment, concepts like rates of unemployment, epidemics or the probability of being victims of a crime are relatively recent developments.

While the two world wars in the 20th century proved severely disruptive to short-term data collection and sharing efforts, the large amount of military investment in intelligence and related information technologies kick-started the post-war drive towards mechanised computing. Investment in information technologies and related infrastructures continued to grow, as did the power of numerical models (such as weather forecast) that enabled number-crunching at a previously unimaginable scale. Research data became well-recognised as political and diplomatic tools. The 'one world' movement towards international cooperation eased efforts towards stabling globalised initiatives of data collection and analysis. Within climate science, the World Meteorological Organization was founded in 1950 to oversee the international linkage of regional weather services, for instance through the institution of a World Weather Watch and the Global Atmospheric Research Program. In 1957–1958, the International Geophysical Year marked both a decisive advance in the commitment of geophysical and oceanographic sciences towards global data exchange, and a diplomatic achievement in fostering good international relations through research communication (Aronova et al.). This meant, once again, a focus on global infrastructures and related institutions.

From the 1970s onwards digital infrastructures for data sharing were being built in virtually every scientific field (Edwards, 2010; Strasser, 2019). There were also efforts to increase global monitoring, which means the tracking of data across many different contexts. The greater availability of computing resources, the growth of expertise and the possibility of sharing data digitally (increasingly over networks) were important factors for this movement towards global data. During this period, the United Nations (UN) consolidated its global environmental monitoring system just as the World Health Organization (WHO) systematised its efforts to map the spread of infectious diseases. The goal shared across these initiatives was the development of tools, such as numerical models, that could help to manipulate data at a previously unimaginable scale. Recall that this is the age of room-sized computers and that computational power was often a limiting factor. There was also a growing exchange of expertise on statistical and computational approaches to data (UN Division for Statistics). Simulations and future scenarios also increased the use and visibility of global data.

During the 1970s, data was increasingly conceptualised as a shareable and reusable asset, rather than something to be collected to be used only once. Data became an object of exchange and reuse. This approach to data was sparked by the cybernetic movement, with its emphasis on modularity and complexity (Pickering, 2011). It was also accelerated by the rising positioning of science and technology as means towards economic growth, military power and international relations. At the same time, Big Science projects carried out at Los Alamos in the United States and the European Organization for Nuclear Research (CERN) in Geneva became a model for how to do research (Price, 1963). Within these programmes, the production and trade of data were no longer the responsibility of individual researchers, but rather the product of large investment and collective efforts carried out in centralised experimental facilities. Even in fields where such centralisation was unfeasible, such as environmental, biological and climate sciences working with observational rather than experimental data, there was a strong focus on building data-sharing networks so as to feed more information to novel computational tools.

The history of data use became ever more tightly intertwined with the history of data infrastructures, and institutions in charge of deciding who should have access to data, what standards and **conventions** should govern what data to collect and how, and how the resulting outputs should be labelled in order to be comparable across time and space (Edwards, 2010; Daston, 2017). This required effective administration and monitoring, a long-term vision of the research domain at hand, and conceptual and technological innovations steeped in specific conceptions of the research objects under investigation – a repertoire of skills, methods, institutions and tools that took decades to develop and continues to evolve to this day (Ankeny and Leonelli, 2016).

Thus already in the 20th century, and even more in the 21st, data became recognised as objects of public interest, particularly by governments wishing to use data to inform policy and the management of commerce, military assets, diplomatic relations and public health. The rise in the social status of data was accompanied by an increasing recognition that methods and logistics of data access and management play a significant role in channelling analysis and, eventually, interpretation.

Many studies have emphasised how data and datafication coupled to digital networks and computational tools have occasioned a societal shift. One of the early concepts used to describe this shift was the **information society** (Bell, 1979). In the information society, information is central to the capitalist system of production, innovation and consumption. The information society is often contrasted with other dominant forms of organisation of society, such as industrial activity. A more recent concept is that of the **knowledge society**. In this more utopian view, society generates, processes, shares and makes knowledge that may be used to improve the human condition available to all its members (Castelfranchi, 2007).

These narratives resonate strongly with the Big Data mythology of knowledge emerging from technical developments in data handling. Without taking anything away from the obvious, enormous impact that statistics, computing and related infrastructures have had on knowledge development over the past century, our brief historical overview emphasises even more strongly how society – and more specifically, the social conditions, motivations and governance of data production, exchange and use – has changed, and continues to change, the status, value and uses of data.

2.4 Conclusion: Data do not speak for themselves

The historical review in the previous section makes clear that what we think of as data, and what we think data are good for, has changed radically over time. Once regarded as stable objects whose scientific significance was determined by a handful of professional interpreters, data are now recognised as reusable goods whose significance depends on the extent to which they are mobilised across a variety of contexts and aggregated with other data, thus growing in volume, variety and value – to the point of driving the very process of discovery. We thus see how the mythology of Big Data is strongly tied to the apparatus of institutions, technologies and economic agreements that effectively enabled data to circulate.

There are powerful forces at work, determining which data are produced, which are circulated and which are used – and how. Data thus do not simply emerge from human encounters with the world. Data are the result of numerous decisions about instruments, the design of data collection, sampling, protocols, statistical tests, categories, scale and granularity. All these decisions are informed by the specific context in which they are made and by the priorities set by the actors. Furthermore, many of these decisions are not necessarily actively made, but are based on tradition, convention, best practices or what is learnt during training. These are not simply biases that can be eliminated: all data creation involves selection. The many material, social, political, institutional, technological and economic reasons why we create data in specific ways explain why 'data do not speak for themselves'.

This has important implications for the Big Data mythology that we reviewed in the section 'Big Data Mythology: Data Transforms Society'. First of all, it makes clear that technology does not rule everything and does not determine social life and data use in any straightforward way. Technology has a fundamental enabling role, and keeps facilitating data exchange and use in ways that were difficult to imagine even 20 years ago: we certainly are in the grip of a digital transformation that is touching every aspect of social life all around the globe (Floridi, 2014). By taking into account the social and economic forces that shape technology development, we can better situate and evaluate the characteristics and effects of this digital transformation, to foster its positive impact and avoid its more damaging effects.

Second, understanding the history of Big Data puts in question the extent to which it is truly comprehensive. Data deemed to be useful for trade by powerful countries have certainly received more attention than any other kind of data. Data about low-income countries, documenting the life of people with little access to computing technologies, are certainly lacking, as are data that are not viewed as valuable by those who have the money and power to produce and buy them. By the same token, data deemed to be sensitive for commercial or military purposes have been jealously guarded among close allies, rather than being freely circulated. Given all this, thinking of Big Data as comprehensive can lead us to overestimate what the data reveals and to underestimate how Big Data, like any other dataset, is selective and exclusionary.

Third, it is important to note how causation still matters in the Big Data world. While reliance on correlations to predict future events is certainly growing, the appetite for explanations of what causes those predictions to come true is also expanding. Big Data cannot, by itself, boost causal understanding of the world. The question is therefore what kinds of knowledge, data collection and data analysis could complement the identification of correlations in Big Data, in order to increase human understanding of how the world works, and why.

This brings us to the fourth point, which is that methods – and specifically methods deployed to compare, evaluate and even critique data and related models – continue to be fundamental to Big Data use, despite the 'messiness' often advocated by Big Data advocates. Knowing whether or not a given dataset is representative of a specific population; being able to train an algorithm on a well-constructed data sample; using contextual knowledge to assess whether a correlation is valid or not – these are all crucial skills in the Big Data world, which call for the exercise of human judgement and cannot be fully replaced by automated tools.

The Big Data mythology strongly underestimates the relevance of social context, the theoretical basis of categories, the importance of accountable methods and the human capacity for assessing complex situations to data work, with serious implications for how priorities are allocated when it comes to management of data. In this chapter, we have shown that data are not an autonomous force that shapes society. To understand the potential of data, big or otherwise, it is best to consider how, why and by whom data as a social phenomenon is considered important.

ADDITIONAL READING

Edwards, P. N. (2010). *A Vast Machine: Computer Models, Climate Data, and the Politics of Global Warming*. Cambridge, MA: MIT Press.

Floridi, L. (2011). *The Philosophy of Information*. Oxford: Oxford University Press.

Hey, T., Tansley, S. and Tolle, K. (eds) (2009). *The Fourth Paradigm: Data-Intensive Scientific Discovery*. Redmond, WA: Microsoft Research.

Porter, T. (1995). *Trust in Numbers: The Pursuit of Objectivity in Science and Public Life*. Princeton, NJ: Princeton University Press.

3

CHARACTERISTICS OF DATA

———————————— Overview of chapter ————————————

Summary

In this chapter, we discuss the characteristics of data and the ways in which they function, focusing particularly on data journeys. This concept is useful to understand the mobility of data: the extent to which their value derives from being passed around and used in a variety of contexts, and what implications such mobility has for how we understand data work. The view of data presented in this chapter builds on the history of data (and particularly Big Data) recounted in the previous chapter, and provides a stepping stone towards the more general understanding of how data fit into the broader knowledge production cycle that we discuss in Chapter 4. The chapter explores how telephone data are shaped by sociological, economic, technological and infrastructural aspects. These interactions all shape the data and make them selective and far from neutral. Several examples in the chapter illustrate the importance of understanding the characteristics of data in order to evaluate and make sensible use of them.

3.1 Introduction: Data do not stay still

One of the important ways in which data can become valuable is through travel and the changes that such travel involves. The value of data as prospective evidence increases the more they travel across sites. This travel makes it possible for people with diverse expertise, interests and skills to probe the data and consider whether they can be useful to answer the questions they are addressing. For instance, genetic data may well be collected as part of a project in molecular biology, and within that context they can be used to study the functions of a specific gene. Once these data are widely disseminated through databases, however, their value multiplies: for instance, clinicians can use the data to investigate the role played by genetics in disease; pharmaceutical companies can use the data to investigate ways to treat patients; and educators can use the data to produce beautiful visualisations of cell biology for use in schools.

To highlight the dynamic and layered creation of data, we now introduce the concept of **data journeys**. Data journeys describe the movement of data from their production site to many other sites where they are processed, mobilised and repurposed (see Figure 3.1). 'Sites' in this definition do not necessarily refer to geographical locations. Sites can encompass diverse times and viewpoints too (Leonelli and Tempini, 2020).

The journey of data across these sites involves several stages. At each stage, work with data takes place. It is this work done in each stage that shapes the extent to which data can travel and become usable for analysis and discovery.

The stages include:

- Data gathering, which could involve the very production of data through measuring instruments (e.g. the generation of location data by satellites) or the collection of pre-existing data (e.g. the retrieval of economic data from an archive).
- Data processing, which includes the procedures used to make the data usable – such as ensuring that the data are in a machine-readable format.
- Data cleaning, through which distinctions between 'data' and 'noise' are drawn, and decisions are made about what features of data to highlight and make machine readable.
- Exploratory data analysis to probe what patterns could be extracted from the data, which often uncovers problems, mistakes or gaps in the dataset and thus sends analysts back to the gathering and cleaning stages.
- Model design, which could include questions of classification (under which label to categorise the data?), description (what part of the world are data documenting?) and fit to specific problems of interest to the analysts (what are we precisely asking?). Again, this stage can require analysts to reconsider their data pool, the ways in which it is processed and the tools chosen to analyse it.
- Visualisation and interpretation, often conceptualised as the end result of all this work.

Interpretative decisions about what the data may eventually be evidence for, and how to prove it, are made throughout the process. Interpretation is therefore

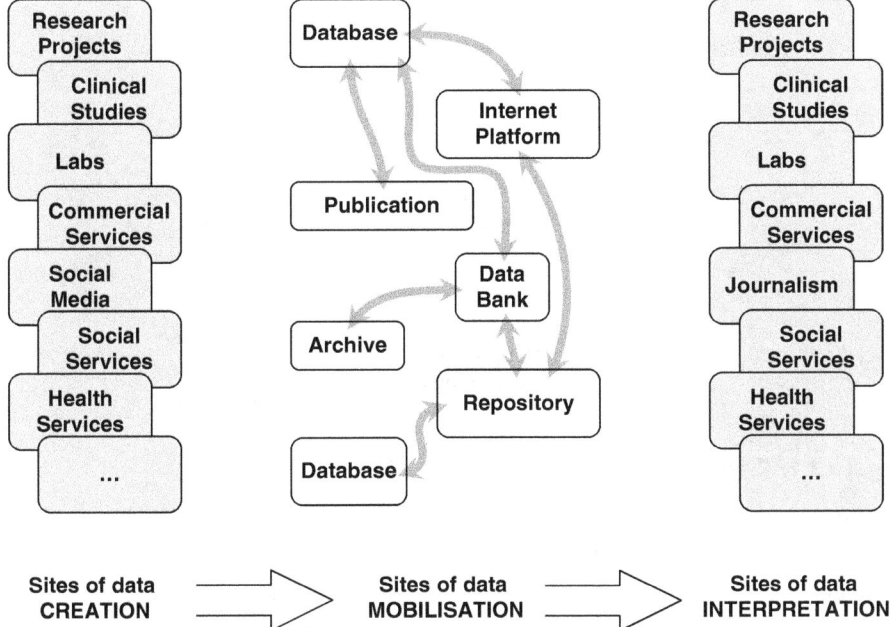

Figure 3.1 The broad dynamic of a data journey, with data shifting from sites of data creation to sites of mobilisation and interpretation

Source: Figure realised by Michel Durinx, copyright held by Sabina Leonelli. First published in Sabina Leonelli (2018) *La Ricerca Scientifica nell'Era dei Big Data*. Rome: Meltemi Editore.

not limited to later stages of data journeys, but is present throughout, even in situations where those busy with cleaning the data do not realise the implications that their choices may have for later analysis. Figure 3.2 illustrates the complex connections and iterations among many of these stages.

Data journeys are anything but linear, and data processing in particular includes many operations that are decisive in determining what comes to count as data, which other data they can be integrated with and how they can be visualised and manipulated as they travel further. Indeed, the metaphor of the 'journey' is powerful because, like many human journeys, data journeys are enabled by infrastructures and actions on the part of humans, to various degrees and are not always, or even frequently, smooth. Data may not be able to travel at all, due to proprietary regimes or ethical concerns. It can also happen that strategies developed to make data travel prove to be unfeasible or problematic as unexpected obstacles and disruptions emerge. This could be because a key digital platform is no longer maintained, because of lack of funding for an essential app or because of difficulties in finding analysts with the

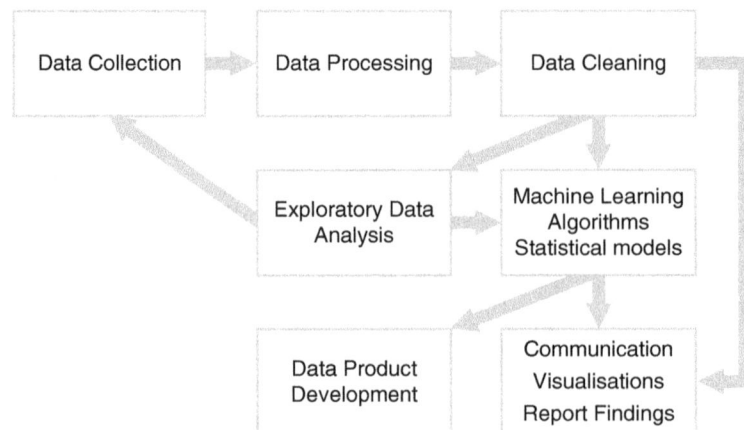

Figure 3.2 Stages of a data journey across sites

Source: Adapted by Beaulieu and Leonelli (realisation by Michel Durinx) from *Doing Data Science* (Schutt and O'Neill, 2013, p.41).

Note that Shutt and O'Neill define data processing as making data machine readable. We use data processing in a different, more general sense, as making data usable. European Union (EU) legislation defines data processing in even more general terms, as doing something with data. Such significant variations are a typical challenge of data work.

appropriate skills. Interest in certain data can change swiftly, especially in highly competitive areas such as biomedical research, where interest in data about a given chemical compound can vary dramatically depending on changing perceptions of its potential value as prospective drug.

Data mobility involves risk, so focusing on data journeys helps to identify and evaluate such risks. To understand risks, we have to be aware of the various destinations that data eventually reach and the ways in which they end up being interpreted (Bates et al., 2016; Leonelli, 2016a; Medina-Perea et al., 2019). Prominent examples of such risks include:

- The emergence of errors in the data. For instance, data are copied inaccurately, or irrelevant noise is included in a given dataset by mistake.
- The loss of data due to careless transfers but also to changes in format and storage, which often make it difficult to use current technologies and tools to circulate and reanalyse data collected in the past.
- The misappropriation of data by people who do not have the skills and/or background information to be able to contextualise them appropriately. This can result in problematic interpretations of the data and unreliable/ false knowledge being generated. An obvious example is the use of data

documenting viewer preferences on streaming services to predict viewers' political affiliation: not only does such a move constitute a potential breach of privacy, but it may also produce false results, since a viewer may choose to watch specific programmes for reasons other than endorsement (e.g. when a right-wing activist watches a left-wing documentary to better understand the opposing side).

- The misalignment between data creation and the uses to which data are put. As we pointed out, data journeys are significant precisely because they bring data into contact with a variety of viewpoints and goals. Where the goals and values of different groups involved in handling data differ considerably, this can give rise to conflict. Consider again a situation where a political party uses data on viewer preferences on streaming services to predict which users may be most susceptible to specific types of political campaigning. There is a potential tension between the interests and goals of the political party in using these data and the interests and goals of viewers when subscribing to a streaming service. Again, this raises concerns around data privacy, which we revisit in Chapters 9 and 10; it also signals the dangers of mobilising data within a very diverse and unequal social landscape, where some actors have more power to act on the data and use them for their own purposes than others.

The focus on data journeys emphasises the contextual and dynamic aspects of data creation and use, and particularly the extent to which data themselves may be transformed by their travels (Leonelli and Tempini, 2020). Transformations may well include physical changes, such as a shift in format or medium (from analogue to digital, from a network to a graph-based visualisation). Regardless of whether data change their format or not, what is always transformed as data travel is their significance as evidence.

Think back to the Data Story on the collection of data on chimpanzees' behaviour in the forest. Data collected on the ground by trackers will be of a variety of types, ranging from numerical counts and samples of chimpanzee droppings to photographs of specimens, weather measurements and observations about the fauna. Once digitised and transferred to a researcher's computer, these data will be used to produce knowledge on animal numbers and movements. But researchers can also choose to post the data on a biodiversity database, where they may be consulted by plant scientists interested in what species of trees are present in the region. Under that new lens, the data will acquire a new significance: rather than functioning as evidence for the understanding and protection of primates, they may be used as evidence for the spread of invasive plant species.

The potential for changes in data use has important repercussions for the data analyst. They need to consider how their techniques and tools are likely to affect the physical properties and evidential value of the data at hand, the speed and ease with which data are circulated and analysed, and the inclusion of certain types of data over others. What types of data best afford which interventions and interpretations? And to what extent do the physical characteristics of data – including their format and the extent to which they fit existing computational tools – constrain possible goals and uses? And what impact do higher or lower speeds of mobilisation have on the reliability of datasets, the amount of uncertainty assigned to them and the extent to which they are reproducible?

All this is very labour intensive and requires insight into the entire process. Lack of investment and strategy around data travel implicitly supports a naive and unrealistic view of data as 'speaking for themselves'. This lack of attention to the journey undergone by data can compromise the extent to which data that travel can be reliably interpreted as evidence. Thus, when thinking of a specific dataset, especially one retrieved from a data bank or digital repository, it is very important to gather information about its provenance and the processing it has undergone. In other words, data should be seen in relation to their journey. Some forms of metadata already exist that help document this (see the section on 'Conventions and Metadata' in Chapter 5). By paying more attention to data journeys, we are better able to question and critically assess how complete, representative and/or relevant the data are with respect to the questions we wish to ask.

3.2 Data are not neutral

Tracking data journeys helps to identify components that are of direct relevance to data use. This helps us to take seriously the way data shape knowledge, rather than dismissing them as lowly building blocks that serve the higher purposes of model building and theory development. This approach is strongly shaped by science and technology studies that stress the way standards, networks and tools shape data, and connect the nitty-gritty of data production to larger issues of politics and knowledge. Data journeys place the spotlight firmly on the complexity of data and the implications that infrastructures – among many other forces, expectations and material settings – have for their interpretation. Such focus provides a strong counterpoint to many of the hyped expectations and unrealistic promises that have come to surround the use of data, and particularly Big Data. Perhaps the most prominent is the idea that research and discovery can now be fully data driven, with novel insights arising from the computational analysis of large datasets without the need for preconceived hypotheses or theoretical inputs. This is sometimes referred to as the 'end-of-theory' claim (Anderson, 2008). As discussed in Chapter 2, hypotheses, categories and theoretical inputs are already part of how we produce data!

Data journeys help us to discover which conceptual judgements and background knowledge are involved. In the current context of data-intensive research, in contrast to more traditional hypothesis-driven research, such judgements and commitments are made at several different stages of the journeys, and often by different people who may not be aware of (or interested in) what others have decided before them, or why. This means that data journey across highly distributed systems, in which many diverse (or even conflicting) perspectives are embedded. Take for instance the moment in which data are generated, such as when taking satellite pictures of a given location or recording athletic performances on fitness watches. At that moment, what counts as data is affected by the specific sensing technologies at hand and the immediate objectives of data collection (which could be the development of geographical maps in the one case, and monitoring the performance of an individual athlete in the other), which determine the frequency, resolution and scope of the measurements in question. Once data are mobilised beyond that context, however, the objectives and constraints of data handling may change. For instance, if health data are shared with an insurance company, the main objective may become to predict the future probability that the athlete may die of a heart attack, while the main constraint would be data protection legislation, which may prohibit specific ways of analysing personal data.

To further explore this issue, we discuss the use of data from telephone communication. The degree of (mobile) phone penetration in many areas of the world is very high by any standard, and it is reasonable to think that data from such practices would tell us a lot. Graphs such as that shown in Figure 3.3 hold the promise of great coverage of data obtained through these technologies. If we compare such rates of penetration (above 100%!) to the response rate of social science surveys for example (usually, the response rate is 12–15%), then it is not surprising that this seems like a wonderfully comprehensive way to obtain data.

A much-used source of data is the call data record (CDR). It is produced from telecommunication transactions, such as a phone call or text message (SMS). Each record has a unique number, and it contains several elements, such as the number called, time of the call, whether it was answered or not and how long the call lasted. For mobile phone calls, cell-tower IDs (base transceiver station) for both caller and callee are also available, which can provide information on location. CDRs are therefore the basis for plentiful and reasonably well-structured data.

These data are in the hands of service providers who use them for billing purposes, to evaluate the quality of their service, and to analyse the behaviour and needs of their customer base. Beyond this primary context, CDRs are also widely

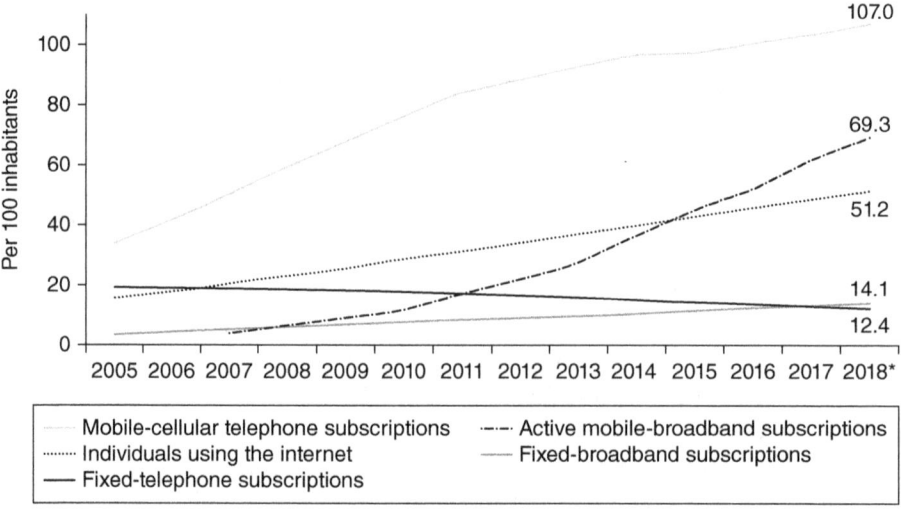

Figure 3.3 Penetration of different information and communication technologies

Source: Reproduced by kind permission of International Telecommunication Union, Measuring the Information Society Report 2018 (International Telecommunication Union, 2018).
*The data for 2018 is a ITU estimate

used to generate data for a range of marketing, humanitarian and scientific research. Mobile and landline phone data have been explored as sources of information on mobility following a disaster, on economic welfare (since higher top-ups are taken to mean more income), on social networks and even on the likelihood of the spread of disease through tracking movement from non-infected to infected areas. Typically, when reporting on the use of such datasets, the main features of the data used in the analysis are described. Exceptionally, authors also report on the collection, cleaning and processing. In the example below, we draw heavily on two texts that respectively describe a dataset used for a hackathon activity (Blondel et al., 2012) and a methodological exploration of network analysis using CRDs (Decuyper et al., 2016). It is telling that these detailed accounts of how data are constituted are not typical scientific journal articles. Published journal articles tend to focus on the analytic methods and results, rather than on data work.

Let's go on a data journey and review what happens to CDRs when they are treated as research data. Typical in the study of social networks using CDR data, many researchers start by removing links or nodes that are not active enough (users who don't make enough calls). Researchers also often remove links that are not reciprocal (if one user calls another user, but is never called), or impose a minimum number of calls to take up a link in the analysis (being called at least six times) (Decuyper et al., 2016). Users who join or leave a provider during the observation period are also usually removed from the dataset. In addition, calls are paired to avoid double count-ing (an incoming call for a callee is an outgoing call for the caller). The time window chosen (days versus months) can also strongly shape the dataset by excluding less active users. Such decisions can orient the analysis in radically different directions. Decuyper and colleagues (2016) have demonstrated that these decisions can go so far as to lead to different distributions in the sample, and that what has emerged as the accepted form of networks may be based on assumptions made at the data cleaning stage rather than reflecting the actual patterns of social networks.

In the study of social networks, data were traditionally generated using ques-tionnaires in which research participants were asked to list their connections. There was always a concern that participants would either provide incomplete lists, because they would forget some of their connections, or that they would provide socially desirable answers – for example, wanting to seem popular and over-reporting contacts, or else not listing some types of contacts if the par-ticipant expected that these might not be socially acceptable. With social media and other telecommunications, there have been claims that radically better data on social networks would become available, since the subjectivity and partiality of the participants' reports would be bypassed. The researcher could capture *all* the relevant data without the introduction of bias from the subject and without missing any of the connections. Clearly, as we discussed above in terms of data cleaning, there is also subjectivity and selection going on with the use of CDRs, albeit of a different kind.

When working with data from mobile phone use, data cleaning can involve a number of substantial decisions about network data. There are other consid-erations in producing location data that also shape the data available. In some situations, the provider may consider that the precise location of the antennas is commercially sensitive information, and so blur the exact location to protect its commercial interests (the position of antennas affects the quality of service delivered, so positioning them strategically is important). At times, the antenna identifiers are simply not available for technical reasons. Location information may also be removed or blurred in a dataset to protect the privacy of indi-viduals: with enough data on calls, the location of one's home or workplace could be deduced by identifying locations from which calls are regularly made. Such decisions about which users to include in the dataset and to what degree of detail location should be included are defensible, and taken for good rea-sons. Yet, it is important to realise that each of these decisions shapes the dataset, and qualifies the promise that we have all the data about everyone.

We now turn to other elements that shape the data. We indicated above that the data are collected by the network operator. This is important for under-standing whose data will be available and what the population contained in the selected sample will be. To what extent are CDRs inclusive? In the context of increasing privatisation of telecommunications and of the growing use of mobile services, which are nearly always privately operated, the users of a given provider are not necessarily representative of a general population. The data provided by an operator reflect who its customers are, and might be skewed in terms of wealth, gender or culture. If the operator is more popular with certain groups, determined by, for example, age, income, gender, language or occupation, the coverage of the dataset will reflect these tendencies. The data from mobile phone use therefore reflect not only calling behaviour but also market dynamics. Furthermore, call activity is increasingly spread over different operators, so that data from a single operator will give partial information. As Decuyper and colleagues note, such biases are very difficult to remove without access to a dataset with a perfect coverage, which does not exist (Decuyper et al., 2016). Indeed, no dataset is ever perfect, and there is no neutral basis from which to generate data. This does not mean we should abandon all hope of 'good data' or that all data are hopelessly biased. It does mean that we can only properly understand outcomes of data analyses if we also take into account the generation of the datasets that underlie analyses and document data journeys.

Besides the selection by researchers and the biases in data production, the behaviour of users can also shape CDRs. Decuyper et al. discuss 'flashing', which consists of letting a relative's phone ring a couple of times, then hanging up and waiting for them to call back. Such a technique ensures that, while either party can take the initiative to have a call, it is always the same person who pays for a communication. But if the filtering technique used to clean the data requires reciprocity (at least one successful phone connection in each direction), then such links will be removed from the dataset (Decuyper et al., 2016). Should such links be retained? It depends on the question. But one may well imagine cases where such relationships, in which one party is willing to foot the financial costs to contact the other party, can be very meaningful social ties and should not be systematically erased from datasets. In addition, it may be the case that phones are shared between individuals, that sim cards get passed around and used in different phones, or that some users have more than one phone (Erikson, 2018) and use different service providers for different kinds of communication (work-related vs personal). All these user behaviours will shape how the data look and what the data can be assumed to mean.

Finally, technological platforms are also determinant of the generation of data. For CDRs, this is manifest in the fact that they do not exist for all calls! If calls are made via Wire, Skype or WhatsApp (or other types of Voice over Internet Protocol (VoIP) phone calls), CDRs are not generated. If particular types of calls are made using VoIP or if specific types of users have strong preferences for these platforms, they will be systematically underrepresented in CDR datasets. Digital platforms also have log files that would enable an analyst to retrieve data about calls made using Skype or WhatsApp. But this dataset would, like CDRs, be shaped by the kinds of uses and users it attracts, and have its own limitations. Digital platforms are furthermore increasingly malleable and generated on the fly – this means that they are less stable as a basis for gathering data. There are always deletions, delays, errors, repetitions, glitches, updates and differences that arise from the many portable technological supports through which platforms are used.

These several considerations about how datasets are shaped even before analyses are performed suggest that we can better see data as a lens, rather than a window. A lens orients us to a particular way of looking at the world, rather than providing a transparent way of looking at it. The assumption of 'having all the data' tends to stress the view that data is a transparent window. As we saw with the example of CDRs, data only document a fragment of reality, seen through a specific perspective and constrained by specific instruments and formats. Second, there are always non-users of a technology or users who use the technology in a radically unexpected way, so there is never 100% representation in the data either. This is neither a shortcoming of CDRs nor of digital data per se. All data sources have limitations and decisions always need to be made about which data to include in an analysis. The problem arises when this need to evaluate, clean, shape and otherwise select data is erased or considered trivial. Furthermore, loud claims about the total capture possible by Big Data overshadow the partiality of data and can cause analysts to neglect making this partiality visible and to underestimate the need to investigate the sources of that partiality.

Returning to the CDR example: the many qualifications made above do not invalidate the use of these data, but they should make us aware of the need to carefully account for the ways in which data are not neutral. As we saw, to evaluate a CDR dataset, we need to be aware of the market share of providers, of the physical characteristics of the transmission system over which calls are made, of privacy regulation, of calling cultures (such as flashing) and of preferences of users for platforms. Sociological, economic, technological and infrastructural aspects all shape the data and make them far from neutral. CDR data can be very valuable, especially when we are able to value them in relation to careful accounts of their non-neutrality. If we take data journeys seriously, then it follows that there is no such thing as neutral data, and that it is not possible to have all the data. The journey of the data has inevitably selected and shaped the data in particular ways. This insight should not be read as a call to remove subjectivity by implementing as much automation as possible or by standardising as many aspects of data creation as possible. What is more insightful and productive is to ask how data creation and the rest of the data journey come into being.

3.3 Data are context-dependent

It is tempting to think that the scientific significance of data lies in their context independence, and the extent to which they objectively document the world without any interference from human interests and values. As shown in the preceding sections, this is, however, clearly not the case. In order to do good data science, one must carefully consider how to contextualise the data as well as the processing tools, questions and background knowledge through which the data are analysed. If we consider data in context, this improves the accuracy and reliability of knowledge being produced, the understanding of the problem and/or situation being studied, and the potential impact of processes of datafication. How data are interpreted often changes depending on the skills, background knowledge and circumstances of the analysts involved, which is why looking at the same dataset from a variety of viewpoints often yields new knowledge. Maintaining an awareness of how data move across contexts, and with which implications, is therefore crucial to the analyst.

Remarkably, studies of data reuse across contexts also show that the expectations and abilities of those handling and mobilising data determine what are regarded as 'data' in the first place (Borgman, 2015; Leonelli, 2016a). Researchers make choices about which of the objects produced through their interactions with the world – be they experimental interventions, observation studies or measurements – deserve the most attention as potential evidence for claims about phenomena or specific courses of action. Biologists, clinicians and plant breeders differ considerably in the data they will consider most useful towards studying gene–environment interactions, and there are many documented cases in archaeology, astronomy, biomedicine and physics where objects considered as data at the start of an investigation no longer have that status by the end of it, or vice versa (Leonelli and Tempini, 2020). A set of photographs taken in a forest, for example, could constitute useful data for the study of phenomena as diverse as the growth pattern of a given tree species, the symptoms of an infection, the effect of certain meteorological conditions on photosynthesis and the presence of parasites in a specific location. Each of these interpretations is affected both by the physical features of the photos (definition, level of detail, focus of attention, colour schemes) and by the manner in which whoever handles these objects accentuates their usability as data (i.e. by zooming in on a specific detail, adding metadata and/or changing format to foster interoperability with other botanical data).

A photo such as the one in Figure 3.4 has a data journey spanning decades. It is a photograph that belongs in the collection of the Tropenmuseum in Amsterdam, The Netherlands. It was probably taken as evidence of the Dutch colonial administrator of the Dutch Moluccas fulfilling his duties. It later became evidence of colonial relations, as it is currently available as part of a

Figure 3.4   Photo as evidence, known in the collection as: The governor of Ceram/Seram, while on a tour through Talseti Bay

Date: around 1940
Location: Maluku Islands (Moluccas), Indonesia
Photographer: Unknown
Source: Tropenmuseum: Collection Nationaal Museum van Wereldculturen. Coll.no. TM-10001632. https://hdl.handle.net/20.500.11840/15282. Licensed under Public Domain Anonymous 70 EU.

digital collection entitled 'Economies and Empire: Colonialism and the Clash of National Visions'. It could also be used as evidence of how software can be used to correct a sloping horizon: in each of these three instances, different aspects of this photograph count as data, and the photograph has a different relationship to technologies and infrastructure (photographic prints, photo albums, colonial archives, digitisation projects, databases of images, image processing software). Within this chapter, this photo is put forward in a textbook, an educational context, as evidence of the importance of data journeys – yet a different use of the photograph as data. Hence, while the features of the objects considered as data certainly shape their use and interpretation, it is often possible to obtain different information from objects depending on how

47

these are managed and interpreted. A particular combination of interests, abilities and accessibility determines what is identified as data in each instance.

Similar considerations apply to data in numerical form. Such data are, on the one hand, eminently transportable: they are easy to copy, aggregate and visualise. On the other hand, what numbers may be taken to represent can vary just as dramatically as in the case of the photograph above. Some numbers may have very different interpretations, such as numbers used to count COVID-19 infections: politicians may see them as documenting the success of a public health intervention, while medical professionals could see them as part of a trend leading to an overwhelmed medical system. Other numbers are taken to indicate different things altogether, depending on who is using them. For instance, the fact that 66% of Greeks had access to a smartphone in 2017 can be seen as indicating a high rate of penetration of digital technology in everyday life in Greece or as indicating that 34% of Greeks have much less access to digital communications. Even numbers that have clear standardised parameters, such as measurements of length, could be interpreted differently depending on which unit of measurement is being considered: it matters a great deal whether a measurement is taken in inches or in metres, for instance.

3.4 Conclusion: Characteristics of data

We started out with some of the claims that data is neutral until analysed and that Big Data is comprehensive. Such a belief in the power of Big Data to provide neutral access to reality and to describe everything has been called **Big Data empiricism**. This indicates a belief that data is the best form of evidence to establish truth, to form opinions about the world and to make judgements: data can somehow 'speak for themselves' and reveal the truth to those who know how to decipher them. Big Data empiricism therefore places Big Data in a privileged position to count as evidence (Rieder and Simon, 2016). This is very significant, especially in combination with the increasing use of algorithms and other automated tools that process such data – something we examine in the next chapter, where we discuss the relationship between data and knowledge.

In this chapter, we gave reasons to mistrust Big Data empiricism. Thinking of data as 'speaking the truth' – independently of how such objects are handled, where, when and by whom – is highly problematic. A much better starting point is to recognise that using data involves being able to contextualise the data appropriately, which in turn involves understanding as well as possible how data have been generated and mobilised – in other words, understanding data not just as a social phenomenon, but as research components. We highlighted the importance of data journeys, and the implications of emphasising the mobility of data to understand the characteristics associated with data. To summarise our findings, here is a list of the key features of data elicited from studying how they are handled and used in society:

- *Data are made, not found*: this draws attention to how data are created, which enables us to better understand how to use them as evidence.
- *Data are partial*: this leads us to ask about what was not captured in the data and about who is excluded. It draws attention to what we expect from users, and to the levels of compliance we assume from users. This also helps us understand how strategies of use (some users avoid using systems in specific ways) or design elements (systems can only be used on some devices) systematically exclude the possibility that some data will be gathered.
- *Data are limited*: this draws our attention to where the data begin and end. For example, we need to pay attention to how changes in legislation, in technology or in cost structures mark the start or end of types of data. Another consideration is the growing and waning popularity of certain platforms among specific demographic groups.
- *Data are shaped by technologies*: we can describe how data are generated and retrieved, and evaluate the kind of variation that can be expected due to instability/dynamism of the systems.
- *Data are contextual, not neutral*: the technologies through which data are generated are themselves products of specific social circumstances, which need to be understood in order to better situate the data.
- *Data contain assumptions*: about what the data could be used for, which were built into the data at the moment of acquisition, and may have been reinforced or changed as data were processed. This makes it important to consider which data were removed in cleaning and why; which corrections were systematically applied and why; and to examine what kind of noise is present among these data and what this noise says about the signal.
- *Data use is contingent on goals*: the interests and aims of data workers play a central role in determining how data are handled and to what effect. As clearly demonstrated by clever statistical manipulations, it is always possible to interpret data to suit a wide variety of agendas. It is therefore imperative to make processes of data analysis and the values and interests underpinning data work accountable.
- *Data are changeable, not fixed*: whenever the format or medium of data travel shifts, for instance when changing the type of file data are encased in to fit a new program, the ways in which data can be analysed and handled also change – often leading to a shift in what data can come to signify.

We need to keep these points in mind whenever using data. Being mindful of the positionality of data fosters the ability to evaluate data according to their provenance, their merits and disadvantages, and to critically assess their value, quality and limitations. This improves the reliability of the data themselves and of the use we make of them as evidence. And in turn, paying attention to the characteristics of data helps us make the conditions, priorities, interests and judgements that shape the data explicit and open to scrutiny.

ADDITIONAL READING

Borgman, C. L. (2015). *Big Data, Little Data, No Data*. Cambridge, MA: MIT Press.

Leonelli, S. and Tempini N. (2020). *Data Journeys in the Sciences*. Basel: Springer International.

O'Neill, C and Schutt, R. (2013). *Doing Data Science: Straight Talk from the Frontline*. Sebastopol, CA: O'Reilly Media.

Strasser, B. (2019). *Collecting Experiments: The Making of Big Data Biology*. Chicago, IL: University of Chicago Press.

4

DATA, EVIDENCE AND KNOWLEDGE

Summary

In this chapter, we build on our analysis of the characteristics of data and examine how data can function as evidence. We show that how we think about data matters. It has important consequences for data management and use of data. We begin by introducing two different ways of conceptualising data, the *representational* view and the *relational* view. We analyse the differences between these two views and then illustrate some of the failings of the representational view – and the reasons why understanding data relationally has a positive impact on all aspects of data work. To discuss the contexts in which these two views are used, we use the term 'knowledge production'. This includes academic research as well as other forms of research grounded on data that produce marketing insights, predictions about future behaviour or indications for running a business. We conclude the chapter with a reflection on how data fit the broader space of knowledge production and explain how knowledge relates to data.

4.1 Introduction: The representational and the relational views on data

When we work with data, whether as researchers in a university setting or as analysts in a company, we have a view on data that informs our decisions and actions. By a 'view', we mean some fundamental expectations about how data relate to the world and how we can rely on them. This chapter is about conceptualising how we think about data and what we think they can do. As an illustration, take the tagline for Twitter: 'it's what's happening'. We might take this at face value – Twitter activity is a representation of what is happening *in the world*, or at least in the lives of the many people who use Twitter. Or we might interpret the tagline differently – this is what is happening *on the Twitter platform*, based on what its users want to communicate about and what is shown to me through my feed as a result of my decisions on who to follow and of the platform's algorithmic selections of which tweets to put in my feed. This very simple example shows the contrast between a representational and a relational view of data.

The representational view of data is perhaps the most intuitive and popular approach to conceptualising data and understanding their role in knowledge production, so we start from there. Within a representational view, data are objects that capture and represent specific aspects of the world. In this view, data constitute the starting point for empirical knowledge. Figure 4.1 provides a graphical explanation of the representational view.

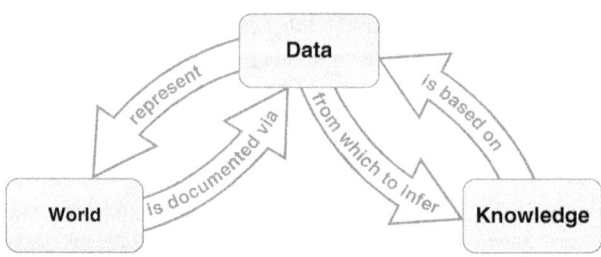

Figure 4.1 A representational view of data

Source: Figure realised by Michel Durinx, copyright held by Sabina Leonelli. First published in Italian in Sabina Leonelli (2018) *La Ricerca Scientifica nell'Era dei Big Data*. Rome: Meltemi Editore.

In the representational view, the more data we have, the more information we have about how the world works – from which knowledge can then be extracted. The representational view acknowledges that when scientists gather data, they do so in a structured way, and that certain aspects of the world are prioritised and highlighted whenever a dataset is produced. For instance, scientists will focus on some aspects and downplay others because of the questions they are asking. They select and structure data on the basis of what they already know and what they wish to know. In turn, this is shaped by historical developments and current circumstances. Because of this selection and structuring, data are not simply a mirror of reality, but rather a way to depict it so that it can be analysed and better understood. Of course, scientists are not the only ones gathering data. When data from social media platforms are gathered, there is also a structuring and selection of data, and some data will be of greater interest than others. In what follows, we use examples ranging from genetics research to the marketing of chocolate.

Raw data, in the representational view, are data that have just been generated and have not been further processed. In this sense, they are the closest documentation that we can have of the world 'as it is', independently of human ideas and interference. Raw data are thus taken to provide unmediated access to reality. This implies that the information content of data is regarded as fixed, regardless of how data are used. The challenge for research methods is then to uncover what it is that the data indicate about the world, by 'cleaning' the data of the noise that may result from imperfect human measuring practices. Controlled conditions of data collection and sophisticated statistical analyses are among the methods that make it possible to evaluate what information a dataset contains about the world. The main focus of research methods is to improve the representations and remove as much human (or other) bias as possible from the process of data analysis. Data, in the representational view, are untainted by human interpretation: the main purpose of data is to help test theories independently of the specific biases and fallibility of human perception.

In the relational view of data, by contrast, the main focus is on understanding data based on the relationships that data have to many aspects of knowledge production, including existing knowledge, social context and human agency. In this view, data are the product of human interactions with the world. They are not representations per se, which have meaning regardless of context. Instead, what data may or may not represent is the outcome of a process of inquiry pursued by humans in a given context. In the relational view, there thus is no such thing as 'raw data' as an abstract category. In interacting with the world, we create objects. What information you think these objects provide about the world (and what should instead be viewed as noise) will depend, among other things, on how you want to use them. The objects will only become data once you have decided that they are to be used as evidence for a particular claim. This intended use is what will help you remove the noise and retain the valuable information. What count as data depends on what you do with them. The question 'What are data?' cannot be answered in the abstract in a relational view. It can only be answered with reference to specific (research) situations, in which investigators make decisions about which data could be used as evidence. This is why it's valuable to define data in terms of their function rather than in terms of intrinsic properties.

When working with a representational view of data, then, the focus is on improving the conditions under which the representation is created – thus, the stage of data collection. Once a dataset is created, it is trusted to contain a nugget of truth that needs to be disclosed through appropriate methods. The bulk of research effort is thus directed at finding good methods (typically statistical tools and/or algorithms) towards analysing the data. By contrast, the relational view defines data as objects that are treated as potential or actual evidence for claims about phenomena in ways that can, at least in principle, be scrutinised and accounted for (Leonelli, 2016a). The relational definition acknowledges that data are powerful but unpredictable objects. Their value as evidence is not fixed and may increase the more data are shared and scrutinised across multiple contexts. The meaning assigned to data, and thus their value as evidence, is determined on the basis of their provenance, their physical features and what these features are taken to represent, as well as the motivations and instruments used to visualise them and to defend specific interpretations. Something becomes data when it is used as evidence for a claim. For instance, a poem can be just a piece of literature; when it is used as evidence to claim that Marlowe is actually the author of Shakespeare's sonnets, that poem becomes data. Similarly, a photograph of a tree can be a nice memory of a sunny holiday, but if it is taken as evidence of the health of that tree on that particular day, that photograph becomes data. These everyday examples are no different from the way data work in Big Data contexts or in audit cultures where indicators dominate. A 'like' on Facebook only becomes data when it is recorded, collected, connected to the liked object and/or the profile of the liker, and compared with other 'likes' in order to serve as evidence to make a claim about popularity, social ties or preferences.

The relational view works with a functional definition of data: data are only such by virtue of their function within a given situation of inquiry, and their relationship to the inquirer, the nature of the phenomenon being investigated and other components (such as relevant models and algorithms). This framework acknowledges that any object can be used as a datum, or stop being used as such, depending on the circumstances. This is a well-known consideration to anyone dealing with historical data, often held in forgotten archives and therefore reduced to meaningless objects. The importance of circumstances for determining whether something is data or not also highlights that the mobility of data matters enormously, as discussed in the previous chapter. The relational view acknowledges that all aspects of knowledge production are connected: changes in one aspect will affect the other steps as well (see Figure 4.2).

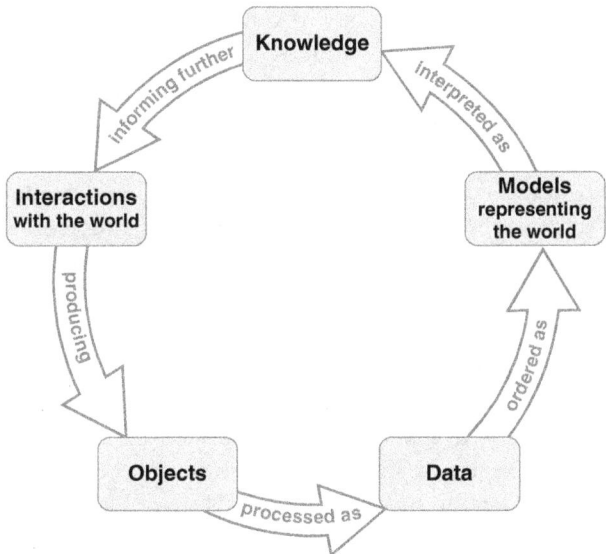

Figure 4.2   The cycle of knowledge production according to the relational view on data

Source: Figure realised by Michel Durinx, copyright held by Sabina Leonelli. First published in Leonelli, S. (2019). What distinguishes data from models? *European Journal for the Philosophy of Science*, 9: 22. https://doi.org/10.1007/s13194-018-0246-0. Reproduced under the Creative Commons CC BY license.

4.2 What is evidence? The path from data to knowledge

What, then, is evidence? In order to understand this better, we need to take a step back and consider how data are actually transformed into knowledge (see Figure 4.1). Empirical investigation starts from the interactions between humans and the world. These interactions produce various types of objects such as numbers, measurements, symbols, photographs, descriptions and graphs. Some of these objects are then selected and processed to become, at least in part, a source of knowledge. What we call data are objects that we manipulate in this way. What these data are used to stand for is not determined exclusively by the physical features of the data themselves. The preconceptions and context of those evaluating their potential meaning also matter. Interpreting data as a source of knowledge therefore goes through another two stages: (1) the development of ways of ordering data that reveal a specific representative function, often understood as the process of data modelling; and (2) the use of the resulting models as an empirical foundation for the production of knowledge. In this view of knowledge production, therefore, the representative function of data continues to be present, but it is not data in themselves that are taken to represent the world. It is instead the **data model**, selected by whoever is interpreting data and deciding how to organise them, that carries out the representative function. In other words, it is a certain ordering of data – the way in which they are viewed and made relevant to a specific dataset – that represents a particular aspect of the world and makes it accessible to further analysis. It is by ordering data that data become usable as proof of specific facts and as source of new knowledge. Data are not by themselves an objective foundation for knowledge; it is the way we organise and view them that determines their meaning.

It is important to note here that the term 'knowledge' can be itself interpreted in at least two ways, and that data are central to how and what we know in both of these senses. The first interpretation of knowledge is as the set of claims that we take to be true, as when 'knowing that' something is the case. For instance, if you want to know what role genes play in the cross-generational transmission of cystic fibrosis, you can look this up in a medical textbook, which will explain this to you. The second interpretation of knowledge is as the skills and strategies that we need to intervene in the world, as when 'knowing how' something can come about. For example, when wishing to know what to do in case of a heart attack, you may sign up for appropriate training by a reanimation specialist. In both of these cases, data are the empirical foundation for the knowledge in question: without data documenting the correlation between the incidence of cystic fibrosis and the possession of certain genetic traits, there would be no evidence for the claim that those genes are a reliable marker for the disease; and without data on the effectiveness of providing certain kinds of help to people suffering of a heart attack, there would be no evidence for the adoption of a certain strategy of intervention rather than another.

Inquiry is thus best depicted as an iterative process consisting of five key steps, represented in Figure 4.2: (1) the production of objects of investigation through interaction with the world; (2) the processing of such objects so that they can function as data, which unavoidably involves a restriction in the evidential space within which data can be credibly used; (3) the ordering of data through data models, so that they can represent specific phenomena; (4) the use of data models to develop knowledge claims about those phenomena; and (5) the use of knowledge claims to frame further interactions with the world. Theory is involved in each of these steps, but in different ways. Steps 1 and 2 affect what ends up counting as data, and theoretical commitments are mostly incorporated in the choice of materials and samples, experimental instruments and data sharing procedures – the ways, in other words, in which researchers carve nature's joints and thus limit the conceptual space within which data can be used as evidence. Steps 3 and 4 are where researchers actively question, identify and stabilise conceptual assumptions about the nature of phenomena to be investigated, which shape the content and formulation of the knowledge claims being produced.

4.3 Examples of data within the knowledge production cycle

Let us consider the example of botanical data to illustrate this process more concretely. In this instance, an amateur taking pictures in the woods produces objects through their interaction with the world. These objects – their photographs – may then be processed by researchers with the expectation they may function as data (e.g. when these photos are formatted and loaded into a database). The researchers order and organise data thus obtained in ways that help them represent different phenomena depending on their interests and specialism: morphologists may analyse the shape of leaves of a specific tree species in a certain location, and use the photographs to create models that represent different leaf shapes and their relationship to the characteristics of different parts of the woods. Or pathologists may look for the visible symptoms of potential tree diseases and use the photographs to produce models that indicate the incidence of a given disease in the forest. The different models created by these researchers are then tested to verify their reliability and relevance to the phenomena they document; for example, pathologists check that the disease symptom model derived from the photographs found online matches the features of data models coming from other sources and, when possible, they return to the location in question to verify the truthfulness of the model. If the models are found to be adequate, they are used as a source of knowledge of the way that disease manifests in the plants analysed. If they are judged to be inadequate, researchers go back to analysing the data and try to order them in different ways – a process that sometimes requires radically changing the type of objects considered as data and/or the aspect of reality being investigated: perhaps the pathologists have been considering the wrong disease, for instance. This example shows how, in order to give rise to knowledge, data need to be manipulated, cleaned and ordered to fit/inspire/support a particular representation of the world, that is, a model. Modelling is the stage of inquiry where a link is made between what are considered to be data and the aspects of the world that the data are supposed to be documenting. Once data are made to fit a specific model, their value as evidence is established.

Another example is the analysis of telephone call data records (CDRs), a type of data that we discussed in relation to data circulation and cleaning in section 3.2. CDRs were designed to keep track of telephone service use and to generate reports that would enable companies to claim funds from their users. In this sense, CDRs were designed to act as data for the telephone service provider, since they were used as evidence of phone use for which a customer could be billed. Beyond this original purpose, CDRs have been used to understand a variety of different phenomena, like mobility and migration, traffic patterns, social networks, market development opportunities, etc. A claim like 'there is demand for more capacity of mobile networks in this part of the city' is a knowledge claim based on objects (CDRs) that become data once they serve as evidence for this claim. It involves using a model of telephone activity and of user behaviour in a physical space that explains the real-world activity that leads to the generation of CDRs. This example shows that such a model is not necessarily explicit, and those who produce it and use it may be doing so without being aware of it. The model may indeed be based on everyday experiences or common sense. If you think 'when people use their phones, this activity becomes visible as a CDR', this is based on a specific understanding of the relationship between phone use and the creation and recording of a CDR. This understanding can work as a data model when it is used to analyse and interpret CDR data. (Of course, a model can also be more conceptual, expressed as a diagram or in abstract mathematical notation – we return to the issue of data modelling more systematically in the next chapter.)

Data are also a factor in other settings, where the traces are produced differently. How about a claim like 'the best time to market our new chocolate bar is when our customers are most likely to buy chocolate-related products, which is between 16.00 and 18.00 local time'? This claim is based on buying patterns that are identified using objects such as timestamps and items on receipts listing chocolate as one of the items purchased. These become data when they are put forth as evidence of purchasing patterns. In their role as data, timestamps and itemised receipts can be compared to check for patterns of purchases, with receipts containing chocolate being clustered depending on time of day. The resulting knowledge about preferred time for buying chocolate will be used to shape future actions, such as prominent placement of ads for chocolate products at given times. In turn, this can be monitored by paying attention to objects that can serve as evidence as to whether the ad campaign is working (e.g. trends in sales revenue).

A final example: consider a claim like 'to ensure that our citizens experience the best service from the municipality, we have to ensure that the passport desk has extra capacity 6 weeks before the school holidays begin'. The claim is based on data about the number of times that citizens access the web interface to make an appointment with the passport office in the weeks preceding the school holidays. The evidence to support this claim consists of the traces left by website use, specifically the 'make an appointment' feature. These data can be combined with other types of data, such as reports produced by employees of the passport office, who identify a large number of 'last-minute' users of their service – thus corroborating the need for higher capacity to ensure that all those who request a passport in the weeks before school holidays can actually get one in time to be able to travel.

Given these examples, we can now come back to the question of what **evidence** consists of in the relational framework. Evidence is the specific arrangement and formatting of data, which is taken to corroborate (provide reasons to believe in) a particular claim. In other words, evidence is 'ordered data': data that have been prepared, managed and visualised within a model to serve as evidence about the world.

4.4 Contrasting the representational and relational perspectives

Both the representational and the relational views on data agree on a number of key points:

- Data are the result of complex interactions between researchers and the world.
- Interfaces such as observational techniques, measurement and registration devices play an important role.
- Rescaling, modification and standardisation of objects are needed to make investigation possible, and this holds for many different kinds of objects, including numbers as well as observations.

However, a relational view draws more attention to the whole cycle of knowledge production. A relational view highlights how the whole cycle is important for how objects underlying data are created, the models that go into shaping this, the kind of mobility we give to data, the kind of knowledge we create and the ways this changes how we act in the world. We saw the importance of this when we looked at datafication and made the link between how this process changes society. If we were to take a representational look at datafication, we would be focusing on the creation and statistical treatment of the traces. Any effects we might observe would then be easy to condemn as 'bad use of data' and we would not pay attention to the way particular data orient us to particular uses (and vice-versa). So, if we want to understand how football metrics are changing how we value players, from a representational view, we would ask: Are we measuring the right behaviours accurately enough? How can we improve sensors? Is the granularity sufficient? Do we need to measure vertical as well as horizontal acceleration to get good data about speed? From a relational view, by contrast, we would ask: What are the effects of the measurements of players becoming evidence of being a good football player, when we use data about completed passes or ball possession or kilometres run during a match? How do these data make sense as evidence, based on our model of what happens during a football match to make a team successful? We might also look at what this does to how we value players (as good/excellent/exceptional), how we train them and trade them, and how we build stadiums that have sensors, screens in the locker rooms and dedicated spaces for the analysts along the field, and how our knowledge of football is changing from an evaluation of teams to an evaluation of players as individuals. We could look at how such data travel to shape FIFA (Fédération Internationale de Football Association) computer games, as well as betting shops and also affect the analytic insights of television commentators.

Another way in which these views matter is in terms of how we value knowledge in our society. Consider the contrast between the relational view and a pyramidal representation of 'data-to-wisdom' often found in knowledge management, systems thinking and management science textbooks (see Figure 4.3).

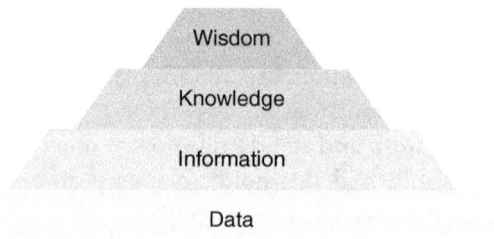

Figure 4.3 Data-to-wisdom pyramid model

Source: Ackoff, R. (1989). From data to wisdom. *Journal of Applied Systems Analysis*, 16: 3–9. Figure adapted by Anne Beaulieu and Sabina Leonelli, and realised by Michel Durinx.

An often-cited source for this hierarchy is T. S. Eliot's *The Rock*, written in 1934 as a commissioned Christian pageant-play, as part of a fund-raising effort to build churches in the suburbs of London. These two lines are usually quoted without much attention to their context: 'Where is the wisdom that we have lost in knowledge? / Where is the knowledge that we have lost in information?' The pyramid model was formalised by R. L. Ackoff (1989). This model supports a number of ways of thinking about the relationships between its levels, and it is generally understood to go from low to high value, and from low to high meaning. Furthermore, when discussed in the context of automation and digitisation, the consensus is that the lower levels can be automated – information might become the realm of AI – but the upper parts may remain uniquely accessible to humans.

This model assumes that there will always be more data than there are claims or evidence, and that from such masses of data some evidence can be distilled, from which, in turn, a narrower set of claims are extracted or inferred. Furthermore, it assumes that data are the least valuable element, while knowledge constitutes the pinnacle of human achievement and is, accordingly, more challenging and valuable. The mechanisms that lead from one level to the next are not articulated, but the assumption is that data are plentiful and form a stable basis for the pyramid, while wisdom is rare. In this model, knowledge is derived from data through inductive reasoning, and data stand as they are, no matter what subsequent interpretation may be attributed to them.

We have shown how knowledge can be much more dynamic than this pyramid model allows: one person's wisdom may be another person's data. Think of a historian studying letters, diaries or reports. While the diaries contained wisdom for the writers, the diaries have the status of data for the historian. Furthermore, we might ask whether there are other components of knowledge and wisdom than such data alone; for example, experiences and relationships. The pyramid thus assumes a representational view of data, in which it is difficult to recognise the key role played by the history of data as a source of knowledge. To those who interpret data as an objective and unchangeable representation of the world, the conditions in which they are manipulated and ordered mean little: what is important is the revelation of their real meaning. This approach matches the idea that putting a lot of data together equates to an automatic increase in the empirical foundations of knowledge. The accumulation of data means the accumulation of a lot of facts; a treasure chest from which to draw new discoveries, via inductive and statistical techniques. It is easy to see how anyone espousing this view is easy prey to the fast promises of Big Data, such as the idea that it is universally reliable, impartial and usable in any type of analysis.

In the relational view, instead, the derivation of knowledge requires that objects selected to act as data (and therefore their physical features) are positioned in relation to other key interpretative features. Aspects that matter include the objective of research, the conceptual foundations and the type of knowledge – theoretical or practical – that is being sought out. This positioning requires deliberate choices and other selections on many fronts. It is not simply a question of which statistical method to apply. The procedures with which data are processed and ordered are critical to their use as a source of knowledge. The relational view of data therefore acknowledges the huge exertion required to document data journeys and makes it possible to scrutinise these during interpretation.

Data therefore do not define evidence, but the other way around: it is the fact that data can be used as evidence that makes them what they are. Similarly, data do not 'contain' information: they are the materials from which meaningful information can be extracted, depending on the circumstances

(Floridi, 2011). Data are best conceived as a relational and not an autonomous aspect of knowledge production. If what is taken to be data changes, other aspects of knowledge production also change – and vice-versa. The types of tools and methods we need, the questions we care to investigate, the types of researchers and users of knowledge can and do change.

How we understand data, knowledge and their relationship also matters when we discuss what is good, reliable knowledge. An immediate consequence of defining data as relational objects is that there cannot be universal ways of measuring data quality and reliability. There is no underestimating the importance of methods for error detection and countering misinformation in contemporary data science, particularly in the wake of the replicability crisis (Mayo, 2018). Nevertheless, most existing approaches are tied to domain-specific estimations of what counts as quality and reliability – and for what purposes. The estimations cannot be easily transferred across fields, and sometimes even across specific cases of data use (Floridi and Illari, 2014). This is a big obstacle to the development of overarching checks for data quality and begs the question of whether producing such context-independent methods is the most useful way to tackle the problem.

Within the relational framework, the reliability of data depends first and foremost on the credibility and rigour of the processes used to produce and analyse them. The unwillingness to acknowledge the epistemic importance of data handling processes translates into an unwillingness to give these processes attention and document them so as to make them visible and open to constructive criticism. The relational view of data encourages care and attention to the history of data, highlighting their continual evolution and sometimes radical alteration as they travel. It also highlights the impact of such changes on the process of extracting knowledge from data.

4.5 Conclusion: Data science in a relational perspective

We can now compare the more abstract cycle of knowledge production to the model of data journeys in data science that we discussed in Chapter 3 (see Figure 4.4).

This combination of the two models enables us to see how data analysis functions as a type of knowledge production. It also helps illustrate the kinds of work needed in different steps of the process of knowledge production. As we turn to the steps involved in data journeys, we will see how these different steps are structured by infrastructures and conventions (among other elements). These make it possible to put data to work. We will also explore the knowledge and skills needed to ensure the integration of these activities, so that the cycle can be completed.

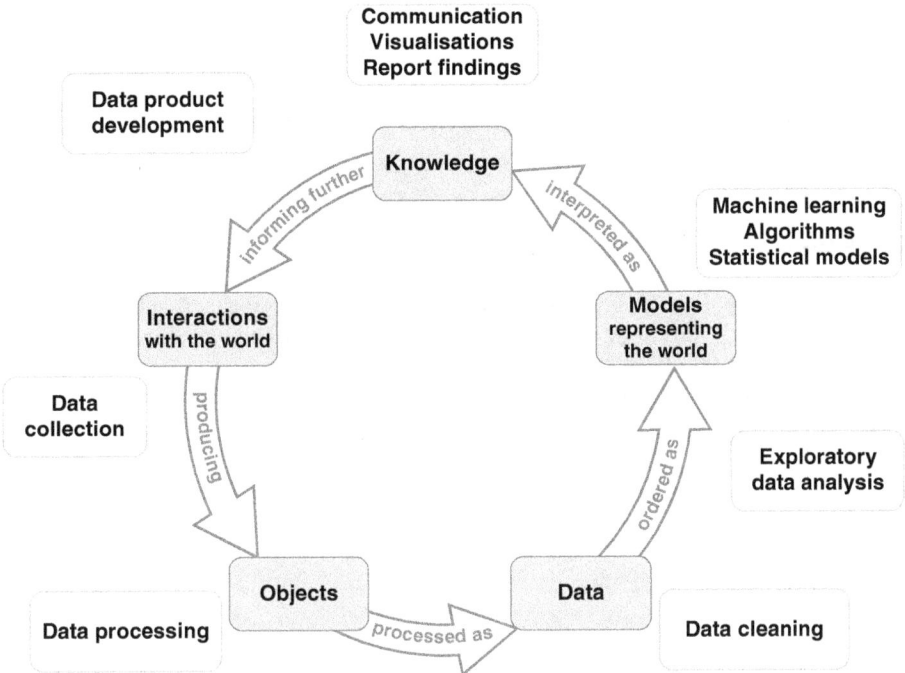

Figure 4.4 The steps of data journeys in data science superimposed on the model of the research process

Source: Adapted by Beaulieu and Leonelli (realisation by Michel Durinx) from *Doing Data Science* (Schutt and O'Neill, 2013, p.41). Reprinted by kind permission of O'Reilly.

We see that datafication is growing, and that data take an increasingly promi-nent place in all kinds of settings. Someone who holds a representational view of data would address this by saying: We have better and better data about the world, more fine-grained, and are capturing more and more aspects of life. This will enable us to have a more rational, modern society because we can base so many more decisions on representations of the world, on real information about the world, rather than on assumptions. We have discussed the limits of these claims in relation to the Big Data mythology outlined in Chapter 2. Someone who holds a relational view of data would say of datafication: We are seeing the entire knowledge system change, including our models (e.g. models of behaviour, with increasingly strong emphasis on individualistic patterns), our interactions with the world (e.g. creating traces and adding value to traces via platforms) and our knowledge (e.g. knowledge that is valued contributes to neoliberal optimisa-tion). We are remaking the world based on our focus on this type of evidence. When we consider the whole knowledge production cycle, we can also see how we are spending a lot of energy on how we package these objects so they can travel, since the focus is not solely on the production of objects.

Datafication happens across different stages of the knowledge production cycle, and what counts as data is affected by infrastructures, conventions, modelling approaches, curation and so on. In the next part of the book, we will explore the broader structures and activities used to make data into accessible and useful sources of evidence.

ADDITIONAL READING

D'Ignazio, C. and Klein, L. F. (2020). *Data Feminism*. Cambridge, MA: MIT Press.

Leonelli, S. (2020). Big data and scientific research. *Stanford Encyclopedia of Philosophy*. Stanford, CA: Stanford University Center for the Study of Language and Information.

Mayo, D. G. and Spanos, A. (eds) (2009). *Error and Inference*. Cambridge: Cambridge University Press.

Radder, H. (2009) The philosophy of scientific experimentation: A review. *Automated Experimentation*, 1(1): 2.

Part III

DATA CIRCULATION

Summary

This part builds further on what we have learned about data journeys. It considers what the contextual nature of data means for how data can travel and explores how data, though not neutral, can be reused, combined or shared. Chapter 5 explains the practices that enable the flow of data, including infrastructures (networks and platforms), **metadata** and related conventions (standards, annotations), models and visualisation tools, and related expertise (**curation**). These all play a significant role in the multiplication of the uses and users of data. Chapter 6 describes the growing place of data science work in and outside academia, the kinds of skills needed to achieve different types of circulation, sharing or reuse of data and how these skills are interdependent. The interdisciplinary work of data science is illustrated using various examples that show the complexity of data science. Chapter 7 examines regimes and governance of data flows: why we expect that some kinds of data can flow and others not, and how this is shaped by values, systems and regulations. Processes of decontextualisation and recontextualisation are also discussed, by revisiting the idea of data journeys put forth in Part I, and considering how looping effects in the data cycle can be implemented. The effects of greater circulation of data and of an increasing diversity of settings in which data is used are examined. The potential of new data flows to support social change or to reinforce of existing inequalities is also set out.

Learning objectives

This part will help you to:

1. formulate data problems and understand the different aspects of data projects;
2. interact with data science specialists and to translate data issues from your own domain to data science;
3. situate your own emerging expertise in relation to the broader data science job market and research landscape;
4. assess and formulate the kinds of work needed for data to flow and for data to be reused, including data handling, data curation, data visualisation and modelling, and data management;
5. identify features of different types of data governance and understand their implications for the re/use of data.

Data Story 3: Geolocation: It's a GIS World

Since 2014, more than half of the world's population lives in cities, according to the United Nations. Increasingly, the way we navigate cities involves digital information. Obvious ways in which we engage with the city digitally are by using Google Maps, Uber and TripAdvisor – all sources of spatial digital information about cities. Many apps also systematically collect and generate location data. Think of a weather app, where having a precise way to measure location is crucial. Many other apps also collect location data, such as Instagram or Twitter, even though location is not immediately related to the primary or explicit function of these apps. As a result, a lot of digital spatial data are generated by geotagging a huge range of digital transactions or activities. Geospatial data can then be combined with data from different sources, such as photos from Instagram, transactions made with a credit card or searching for a dining recommendation. Geospatial data are linked to everyday actions of individuals, objects and processes, a process that we characterised as datafication.

- *How can such different data sources be combined in relation to specific places?* How do these data come together in a geographical space? How does this enrich the profiles that companies are able to build and how does it provide valuable data to help predict behaviour?

Underlying technologies of geolocation make it possible to combine, connect and correlate the data across these many sources. Location data therefore often serve as a reference point to integrate diverse datasets, since it is typically assumed that location measurements are fixed and objective regardless of the instruments that produce them and the context in which they are used.

- *How do diverse settings and tools come to share infrastructures and standards?* What is the motivation to collaborate and trade data? Which actors have to work together to achieve this?

The Global Positioning System (GPS) uses about 30 satellites and is owned by the government of the United States. It has close ties to the military and is operated by a branch of the American armed forces. The satellites transmit signals to electronic receivers on Earth, so that these receivers can determine their location relative to the satellites' positions. The GPS system connects to a reference system that links position with location, currently the *World Geodetic System* (WGS 84). This requires complex calculations based on data from the different elements in the system, and very precise calibration.

These connected systems are American-funded, US-based endeavours, have global reach and can be used free of charge. But not all countries want to depend on this American suite of standards and technologies, which is seen as imperialistic in many parts of the world as well as a threat to national security due to the fine-grained accuracy with which it fosters the identification of specific locations. The system is sometimes embraced and sometimes resisted. The Chinese government for instance has developed its own alternative geographical system called GCJ-02. This system is known to use an algorithm that adds random offsets to the latitude and longitude provided by the WGS-84 system. This means that a location produced within this system is inherently imprecise, making it impossible to use the system to generate high-definition measurements. Thus, a building in central Beijing may be located in the middle of a lake, even when this is obviously impossible: in reality, the building will be located close to the lake, but the system does not allow its precise location to be pinpointed.

- *What implications is this strategy likely to have?* Does this algorithm improve Chinese national security, and how? Should other countries also develop a national system for location measurement? And if many such systems emerge, which standard should be used for international communications and services?
- *What does this case tell us about location data?* In which sense are measurements of location neutral and context-independent? Is it possible to use location data without considering the differences between georeferencing systems and their political and cultural significance? Can you imagine situations in which data workers would need to consider these differences in order to produce reliable insights?

It is important to note that even within a single georeferencing system, we can expect, given the material discussed in Part II, that geolocated data are not uniform and objective. With regard to geographical data, there are differences due to the

(Continued)

economic value attached to certain locations or to the affluence of particular groups who can afford digital devices and connectivity. Some areas may be overrepresented because of their importance for the map producers – Google for instance is more interested in mapping commercially dense areas than deserted landscapes, because these spaces are more relevant to Google's customers. There are also significant absences, even when data are not produced by corporate actors like Google. Crowdsourcing of geodata in OpenStreetMap reveals that the tagged spaces reflect the gender imbalance of its contributors. For example, the classification of bars and sex clubs is presented with more granularity than types of childcare facilities – the categories that matter to the taggers are included in the maps.

- *Does it matter that not everyone sees the same map?* How do different dynamics of personalisation filter and order spatial information we are presented with, and to what extent? How does this digital geospatial information shape how we experience space, how we navigate that space, and how space is planned and organised?

Another element to consider with regard to geolocated data is how ubiquitous some types of maps have become. For example, we now expect that we will be able to toggle between satellite and map view. This actually depends on the possibility of data integration and data flow. We also expect to see a route overlaid onto the geographical space, depending on our mode of transportation. Yet, there is nothing obvious or natural about such interfaces – these get overlooked because once we accept them as conventions, they seem transparent.

- *Why do we take such interfaces with data for granted?* Is it possible and advisable to consult more than one type of map, or can we simply trust the map we use every day on our smartphones? Which skills have we learned to be able to understand and act on these kinds of maps?

Data Story based on Stephens, 2013; Kitchin and Dodge, 2011; Brunton and Nissenbaum, 2015; Shaw and Graham, 2017.

Data Story 4: Tracking Tuberculosis Using Phone Data

The 20th century has seen major advances in the fight against infectious diseases. So much so that it seems that we might now be able control their spread, since they tend to be highly local. One of the first reactions to the emergence of the COVID-19 pandemic, for instance, was to set up systems for tracking the disease using mobile phone technology. This approach builds on the processes of datafication that we have discussed earlier. It also relies on interactions between different types of data, experts and users.

Major efforts to track disease have focused on tuberculosis (TB), a complex infectious disease that is still endemic in many countries in the world (a situation made even worse by the COVID-19 outbreak). Many people can be infected with the disease without developing any symptoms, but for those who do develop symptoms, TB is deadly in about 50% of cases.

TB can be treated with a 6-month course of antibiotics. The bacteria that cause TB are carried in airborne particles that are generated when a person coughs, sneezes, shouts or sings. Transmission occurs when a person inhales these particles, and the bacteria reach the lungs. Given this infection mechanism, the frequency and duration of exposure to an infected person are two major factors governing the transmission of TB. For this reason, knowing about the contacts of a patient suffering from TB is important, since these contacts may themselves have been infected.

An attempt to map the spread of tuberculosis in India was part of the activities of the Big Data for Social Good Initiative, a public–private partnership launched in 2017 aiming to contribute to the Sustainable Development Goals of the United Nations. The government of India was already involved in different public health initiatives as part of Be He@lthy, Be Mobile (BHBM), using SMS in different campaigns. But the BHBM consortium thought that mobile phones could be used in a more innovative way. In collaboration with mobile phone operators, the World Health Organization (WHO), the International Telecommunications Union and Airtel pursued a pilot project in the Indian states of Uttar Pradesh and Gujarat, India being one of the countries most affected, with one quarter of the total number of deaths globally (WHO, 2019).

This project used mobile phone network data in combination with publicly available data about incidence rates of TB for different areas (the incidence rate is the measure of the frequency with which a disease occurs – how many people get sick with TB every year). The incidence rate of TB per region was combined with movement patterns of about 40 million mobile phone users as derived from mobile phone network data. The analysis showed that when an area with few cases of TB had high mobility (e.g. through people commuting between home and workplace) to areas with many cases of TB, the low incidence area was at risk of increasing TB levels or could already be underreporting TB cases.

Understanding these patterns could make it possible to act, for example by implementing vaccination programmes and awareness campaigns or by deploying additional clinics in the affected area.

(Continued)

- *What do data workers employed in such a project need to know?* Is it relevant to know something about TB in order to provide reliable results? What kinds of expertise are required to make some sense of these data?

In order to understand how mobile phone network data could reveal something about movement of users, the analysts need a good understanding of how mobile phones connect to antennas/transmission towers, of the granularity of these data and their variations in space (in some areas, there are fewer towers, for example), and of the kinds of noise that could be expected and would need to be removed from the dataset. Hence it could be argued that analysts need to know the characteristics of the areas under study, or work with people who do (such as urban planners or social scientists).

To know which kinds of data analysis would be relevant for understanding the spread of TB, medical expertise is also needed. With TB, repeated exposure is an important factor, which is why regular patterns in movement, like commuting, were taken into account, rather than the absolute number of movements. Known risk factors for TB, such as overcrowded housing, medical malpractice and poor awareness of the symptoms among patients, also need to be taken into account (Dye, 2014). Last but not least, knowledge about public health and local healthcare requirements is also necessary. For example, if healthcare is organised at the municipal level, state-level data will not correspond to a logical level of action for those who make decisions about healthcare. In the TB project, these many capabilities were brought together in a process of joint learning and iteration.

- *How can such diverse sources of expertise be integrated in such a project?* What do experts need to share, and in which forms and venues? Is it enough to share data? Do experts need to explain to others why particular parameters for analysis are important? Does explaining slow things down, and is this a problem? How can experts appreciate the requirements and skills brought to the table by others? How can they tell the difference between an important requirement and a shortsighted demand from someone who has not understood the problem?

A final consideration is the kind of regime of which these data were part. The project was a public–private partnership.

- *Who are the actors involved and what is their stake?* Do the relationships between the Indian government and Airtel matter for the results of the project? Who is dependent on whom? If Airtel considers that the dataset contains commercially sensitive information and does not wish these data to circulate, how can the findings be verified by other scientists? And why shouldn't others benefit, if the WHO and the Indian government, both public organisations, have made major investments in this dataset?

We can also consider what happens in the aftermath of such a project. This seems to have been a one-off intervention. It is conceivable that it might have longer-term consequences:

- *What can we infer from such a project?* What if the local health authorities start to organise their work according to the findings based on the algorithms developed in this project, using it to plan where mobile clinics will be set up? Is Airtel obligated to keep providing data? Does state healthcare become dependent on data philanthropy from corporations?

Such a project might also be pursued at different scales. For example, going from large-scale to more targeted identification of TB transmission within small geographical areas.

- *Who is targeted?* What if this project were further developed, and shifted from identifying patterns in the population to identifying individuals? What would happen if rather than using anonymised data at a large scale to find general patterns in the population, mobile phone data were used to track individuals (as was suggested in the early stages of the COVID-19 pandemic)?

Such developments would imply linking phone data to one's health status. Such data are usually considered to be highly personal and private, and access to this information tends to be highly regulated.

- *Who should be allowed to access these data, and under which circumstances?* Would we expect personal data, such as being diagnosed with TB to be available to actors like Airtel, so that these data can be combined with movements as derived from mobile phone data? Would it be acceptable for Airtel to work with such data?

If these data analyses could lead to monitoring movements at an individual level, the quality of the data would matter a lot.

- *What could be the consequences of errors in the data?* How could the project team ensure that the data are of sufficient quality, so that individuals would not be unduly penalised due to errors in mobile phone data? And could the authorities demand that citizens use an Airtel service, in order to monitor their movements and maintain public health?

Data Story based on Dye, 2014; Fleming et al., 2017; GSMA, 2018; Beaulieu, 2021.

5

PUTTING DATA TO WORK

Overview of chapter

Summary

This chapter identifies and discusses technological, infrastructural and practical elements needed to put data to work, and particularly the 'messy' data produced by human digital activities. We describe networks and platforms used as infrastructures to store, structure and share data; standards, conventions and metadata; models and visualisations, including infographics and other specific ways of clustering data to extract meaning. And, last but not least, we also explain the curation practices through which data are maintained and cared for, and without which data would not be accessible and usable as evidence.

5.1 Introduction: The complexity of putting data to work

Since the second decade of this century, data has been at the centre of promises to revolutionise all kinds of sectors and to provide huge benefits to society. These promises have focused on Big Data and often involve scenarios of merging data flows. The hope is that, starting from globalised data collections, we can combine datasets, mine them to find patterns that would otherwise go undetected and use these patterns to deliver better services, support development, and create new kinds of businesses and activities. The more areas of life become infused with data through datafication, the more potential there seems to be for data to play an important role. The analysis of social data gathered from large internet platforms, social services and more traditional industries is widely expected to inform evidence-based policy, business strategies and education, possibly even replacing traditional data production in social sciences. All these changes also highlight the importance of data science as the field where statistics, maths and AI can be applied to such data riches – a question we will explore in the next chapter.

Clearly, such processes involve not only data, but entire suites of methods and technologies. Most notably since the advent of portable computers and devices, data collection and archives have grown dependent on digital technologies. They are deeply interconnected with the rise of computational modelling and simulations. These technologies include hardware, for example, the use of multiple computers to store data and solve problems in a distributed way. Methods from AI are also needed to process the data input, such as machine learning and deep neural networks. Further tools for **data visualisation** are indispensable. There are also technologies that we can qualify as social, in the sense that they function mainly based on practices and conventions rather than on physical properties of machines (Derksen and Beaulieu, 2011). All these technologies function together and constitute potentially powerful assemblages that make Big Data work towards goals such as personalised medicine and precision agriculture. These also involve the dedicated work of many kinds of experts who have developed new ways of working, repairing and caring for data. Many institutions have adapted their organisation and work flows to ensure that the work needed to make data available and useful can get done. As a society, and particularly after being forced to rely on digital technologies during the coronavirus pandemic, we have also learned to engage with data, becoming adept at decoding data visualisations, infographics or other interfaces for everyday or specialised purposes. In this chapter, we identify key elements needed to put data to work – technological, infrastructural and practical.

5.2 The challenge of 'messy' data

To understand how data can be put to work, we first consider two different dynamics that shape contemporary digital data. On the one hand, global systems have been painstakingly developed, often based on over a century of analogue data collection and involving complex negotiations among actors around the world. Typical of such global systems are the data collecting and processing practices of the World Health Organization (WHO, 2019), which monitors specific diseases around the world based on data generated according to specified protocols that yield standardised data, and other similar international agencies. National organisations such as statistics offices are also important actors in creating data and making it widely available for administration or policy purposes. On the other hand, there are more recent developments, where data is the result of very diverse activities – transactions, interactions, communication – rather than the result of deliberate observations or measurements. In this second dynamic, data production and collection are much less regimented. 'Messy data' abounds as the digital traces of online human activity, and it is the size and the possibility of combinations that makes data valuable. We briefly consider these two dynamics, before moving on to the analysis of how data is put to work.

The first way in which data has become increasingly prominent is through the growth of global systems. In many areas, standardising and monitoring systems of measurement and data collection is now a priority, and the power of institutions tasked with these goals increased accordingly. Climate scientists have developed sophisticated ways to use legacy data to reconstruct a history of the atmosphere at the global level, including models to bring together climate and weather data – and this effort in turn fostered further efforts towards the pooling of international data, culminating in the 1992 establishment of the Global Climate Observing System (Edwards, 2010). In biology, the quest to map biodiversity moved to the molecular level with the start of big sequencing projects, first in model organisms such as the worm *Caenorhabditis elegans*, then through the Human Genome Project (Hilgartner, 2017). Sequencing databases were reimagined as environments for in silico discovery that would facilitate immediate, low-cost data-sharing, visualisation and analysis via the internet, thus helping to transform the massive investment in genomic data production into useful biological and medical knowledge. Many other areas have developed such systems of standardised data collection, for example with regard to biodiversity, pandemics or to support progress on the United Nations' Sustainable Development Goals (Beaulieu, 2021).

The second dynamic seems at first much less ordered and may appear as the result of organic growth rather than concerted effort. Here, we can think of using Twitter data to assess 'public sentiment' or to predict the likelihood that an employee will start looking for a new job through patterns of activity on a site like LinkedIn. Data gathering can therefore seem to be much less organised in these contexts, and to be more valid because it is generated 'spontaneously', rather than through the deliberate design of a scientific experiment or field trip. There are many cases in contemporary society where data that was generated for no explicit purpose can end up being used for seemingly concrete ends, including policy decisions. For example, data from FourSquare (where users check in to a particular location) in combination with Twitter data has been used to quantify the degrees of diversity and homogeneity of neighbourhoods in London. The data was used to make claims about whether some neighbourhoods had a lot of visitors (high social diversity) or were populated by a fairly constant group of people (low social diversity). These 'social metrics' were correlated with other indicators for wellbeing. The researchers found that signs of gentrification, such as rising housing prices and lower crime rates, were the strongest in deprived areas with high social diversity (Hristova et al., 2016). This innovative use of social media to address classic questions in urban development and policy is exemplary for how messy, seemingly spontaneous data is put to work, to contribute to knowledge and policy.

At the same time, both dynamics – whether they seem scientifically ordered or spontaneous and messy – rely on a set of practices that are essential to putting data to work. These structuring practices involve:

- a high degree of automation in data collection;
- increased data storage, processing and analysis;
- the production of copious amounts of metadata that document the provenance of the data and can themselves serve as useful data depending on future uses;
- the mobility and interoperability fostered by the development of data infrastructures and related analytic tools.

While many of these developments feel intimately tied to the digital, they also have roots in the histories of bureaucracy, standardisation and engineering. As we zoom in on these various elements, we will note why these roots matter and how datasets need to be constantly maintained and repaired by humans. The digital sphere is both dynamic and multiple and can feel endless in its capabilities and potential. Yet the digital sphere is structured and constrained by series of tools, conventions and practices that support particular kinds of interactions and values. These elements make it possible to coordinate activities, whether scientific, social or commercial, across global regions. These structuring elements are largely invisible, yet they have become central to global capitalism and to cultural forms – effectively defining life in the 'information age' (Castells, 1996).

Why are structuring elements important? To travel by car, we need an infrastructure that takes the shape of a network of roads and highways. We need arrangements for fuelling our vehicles. We also need stop signs and traffic lights, and these function by virtue of a series of conventions (red means stop) that can be enforced (driving above the speed limit can lead to being stopped by a police patrol car). Furthermore, we need to be able to find our way and interact with the road network – think about maps and Global Positioning System (GPS) navigation systems. Roads enable us to go places by car; they also direct us towards particular destinations, since highways make journeys more likely, and discourage us from visiting certain places by car if no suitable road leads there. Data circulation and use relies on similar, layered suites of technologies, and these suites of technologies orient us to particular practices. If we understand how data infrastructures, conventions, interfaces and curation are organised, then we can better understand why data is used in specific ways and not others. In this chapter, we zoom in on different structuring elements and on the practices that make it possible to put data to work.

5.3 Infrastructures

In this section, we consider series of infrastructures that enable us to put data to work. Typically, infrastructures are reliable. They are everywhere and tend to remain invisible until they break down – then we sorely miss them and realise how important they are in structuring our activities. For example, at our place of work, which is the university campus, we are so used to using computers for research, teaching, communicating or studying that when there is a power outage, we are not able to get much done. Or even when the internet is down at work, we find that our access to our documents, sources, data and colleagues also breaks down because cloud storage creates a dependency on this network. Networks and platforms are two key forms of infrastructure for data work.

5.3.1 Networks

In the data story on fighting tuberculosis, the very possibility of following mobile phones depends on the presence of a network of towers that provide signal to the phones and therefore track the devices in order to connect them. How did we get to such a networked situation?

Networks that can be used to transmit digital content have become ubiquitous, linking billions of devices that are built to produce, store, transmit and handle digital traces that can become data. These networks have their origin in organisations with a public mission (government, military, research) but have become increasingly corporate following the privatisation of the internet in the early 1990s. The early stages of the commercial internet were shaped by the practices and structures of telecommunications, so that telecom providers and system operators often took on the provision of internet services.

Electronic networks make greater circulation possible, which means that the site of creation of digital traces may be far removed from the site of use of data. Traces created in a certain time and place can end up being used in a completely different location and in a different context. Networks should be understood as the elements that enable connectivity: they are not 'simply' connected wires, but also include the technologies that enable wireless connections. Such 'networked ICTs' are a combination of hardware and code (Postigo and O'Donnell, 2016). As we enter the third decade of the 21st century, an estimated 3 out of 7 billion people in the world have access to ubiquitous computing (Zuboff, 2019). Circulation and connectivity rely on an infrastructure that makes it possible to sense, record, transmit and process data.

The development of networks is not only pushed by Big Tech – there is also a 'demand' side to the dynamic of growing networks. There are strong associations between connectivity and self-realisation, as well as between networks and development. Many activities that contribute to our identity and sense of self involve the use of networks – whether to maintain social connections or to pursue our passions. When understanding our health, staying in touch with our colleagues or finding true love depend on logging on to the internet, technology and our daily practices are entwined. This connection between who we are and networks makes them unmissable. Furthermore, many of our more collective aspirations, for example, for sustainable development or decreasing inequality, have become associated with the roll out of networks: providing access to the internet is increasingly considered a pillar of regional economic development or of access to services. Initiatives to combat the 'digital divide' or campaigns that posit internet access as a fundamental right are evidence of the association between networks and the ways of life to which we aspire.

5.3.2 Platforms

Networks are one of the enabling conditions for the rise of **platforms**, the second main aspect of infrastructure we will discuss. Networks and their effect are what enable platforms like Facebook to grow so rapidly (Srnicek, 2016). Platforms are reliant on the connectivity of potential users. This explains why companies like Facebook are eager to provide access to the internet. This can mean supporting technological innovations that would enhance connectivity and circulation, such as balloons carrying internet capacity, or investment strategies, such as Google outbidding other providers for free WiFi of Starbucks locations, or corporations pursuing laptop philanthropy.

In Chapter 1, we noted the importance of platforms for datafication. In this section, we focus on the role of platforms in structuring data work. In the first instance, platforms can be seen as enabling interaction and transaction; for example, in the case of FoodDrop or Deliveroo, the platform connects restaurants and customers, enabling one party to advertise its menus, and hungry people to browse through a range of options. The platform facilitates the payment and coordination of orders. As such, it acts like a marketplace. However, this is only part of what platforms do. We will look in more detail at the kinds of marketplaces that platforms shape and at platforms as business models in Chapter 8. For now, we can focus on the other activities that are enabled by platforms and what make platforms a distinctive form of infrastructure.

Platforms connect different actors, provide access to data through application programming interfaces (**APIs**) and foster further development of functionalities. APIs are a kind of interface that makes it simpler to connect applications to existing platforms. APIs enable data to move between different software in a coordinated manner. One way to think of it is as a 'socket' in which to plug in. The connection is made without having to worry about how the entire house is wired and how electricity is provided. APIs therefore invite connection to the platform, but also shape how this connection can be made (again, think of a socket requiring a specific shape of plug).

Platforms can be defined as a programmable infrastructure upon which other software can be developed and run (Gillespie, 2017). Platforms are distinctive because they are generative and open-ended. This means that they support the development of further possibilities for interactions by providing data, and that they benefit from the increased data generation from increased interactions. Platforms have two important functions: they are the site of data generation and combination (datafication) *and* support the development of applications (programmability). An example of this is the way Facebook can interface with your email account to find 'friends' on Facebook. Facebook has a wealth of data about all kinds of users and builds complex profiles from each account (datafication). The program makes it possible to search email addresses, parse them and attempt to associate them with profiles and specific accounts on Facebook (programmability). If you use this program, you are further contributing to the process of datafication, enabling the platform to make even more connections and enrich profiles, while you re-create your social network on this platform (this is what we mean by generative and open-ended).

Typical of platforms is therefore that they welcome certain types of interactions to further develop their services. They are meant to enable others to innovate and create new services and new connections. At the same time, the further production of digital traces is integral to their set-up: an app that only provides a service is not as interesting as an app that also engages users to produce more digital traces. This combination of providing a site that supports innovation and creativity and while tying developers and users to the platform and ensuring that further data is generated *on* that platform means that the platform organises labour and captures data as a form of input for profit for the platform-makers (Gillespie, 2010; Plantin et al., 2018).

Social media platforms are familiar instances. On these platforms, there is significant generation of data, because the platform facilitates the production of content by users (photos, comments, messages), the systematic collection of traces (timing of interactions, downloads, speed of typing, etc.) and the extraction of interactions (liking, sharing). Typically, new services are developed on a platform, enabling the delivery of further services or functionalities (for which users may want to pay or that may serve as a support for advertising). These in turn engage users to increase their activities on the platform and to generate increasing amounts of data of different types. Through the combination of different types of datasets, further analyses can be done.

Platforms also organise user participation and therefore everyday life. Platforms make participation possible along specific types of activity, such as sharing, following or tagging, and therefore shape user interaction (Alaimo and Kallinikos, 2017). Participation and interaction generate traces of user activity that can be analysed. These analyses yield profiles or support predictions of user behaviour – or even nudging of user behaviour. Many media outlets have recently featured discussions of 'bubbles' and echo chambers, which are specific instances of how platforms structure exposure to news. Platforms foreground some information and background other content in ways that are not transparent to users and that may shape public life in unaccountable ways.

Platforms are also open-ended. But this openness is not limitless. What boundaries affect platforms and how they function? On the one hand, it is difficult to predict how users will precisely engage with the platform and what kinds of new uses will emerge from the efforts of developers using APIs. On the other hand, there is also a kind of lock-in that ties the users, developers and data to the platform they are using, thereby ensuring that the newly generated data flows through the platform and can be harvested by the platform owners.

APIs can be seen as ways to align content developers and platforms: they constrain what actors can do. A typical example is the growing use of authentication via a Facebook or Google profile for other services. By connecting to new services using these profiles, the data you generate in a new setting is tied to your Facebook or Google identity. Each time a Facebook identity is used for a new service or application, these 'silently' contribute to the Facebook social graph via the API, extracting data from your shopping habits or information-seeking behaviour and sending it back to Facebook. Facebook is then able to use data to personalise advertising and newsfeed or otherwise customise your experience of the platform (Plantin et al., 2018). Similarly, when we speak of 'mash ups', we are dealing with the possibilities for data integration provided by an API. Plantin et al. put forth the example of Google Maps, where an API was released very early on after the launch in 2005. The API enabled third parties to add or overlay data onto the Google map. These are effectively 'mash ups' that take Google Maps as a base and transform Google Maps into a platform (Plantin et al., 2018).

Platforms often claim to be neutral (Gillespie, 2010), but if we analyse them as infrastructures, we see how they shape digital spaces and their users (participation) as well as who can benefit from them by becoming the owner of data and monetising their value. Platforms also have requirements about who can join and on what basis. These requirements become visible when there are debates about 'fake accounts' on Twitter, or on the exclusion of 'bots' from social media platforms, a limitation that may be harming research projects.

Finally, platforms are tied to business models and specific ways of making money. Quoting Hal Varian, Google's long-time chief economist, who speaks of 'data extraction and analysis' as the core of 'Big Data', Zuboff identifies practices that turn the harvested data into input for the design of modes of prediction that benefit commercial interests. Most of this harvested data is neither personal data (like our address or date of birth) nor explicitly generated by users (actively clicking a 'like' button). Rather, it is data about time of use, length of sessions, hovering of a mouse over particular items and other automatically generated microdata. On the basis of this seemingly worthless data, new predictive tools are developed that form the basis of surveillance capitalism (Zuboff, 2019) or platform capitalism (Srnicek, 2016). This is what we mean when we say that platforms are generative: more interaction means more data, which further feeds the growth of platforms' profit and influence. Generation of data for combination for added value is the logic around which these platforms are built – an issue we will return to in Chapter 8.

We have singled out networks and platforms for their structuring roles in providing indispensable arrangements of tools and programs that enable data to be put to work. Networks make the access to platforms possible, and platforms make it possible to access new domains of human behaviour as people increasingly mediate their lives, and to shape the public sphere as well as how we navigate our world. While this section focused on infrastructure because it is often less visible or noticeable, we should not overlook the fact that many other technologies, big and small, are also significant – whether hardware, from giant server farms to tiny wearables like running shoe pods, or software like Haddoop's data management system or apps for smartphones.

5.4 Conventions and metadata

So far, we have stressed repeatedly how data does not stand on its own but is part of a rich and layered context of production, transformation and use. For data to come into existence, to be transformed and to be analysed in order to support knowledge claims, a lot needs to be in place and much work has to be done. We have also amply illustrated in the data story so far that digital data are transformed across suites of technologies (Shove et al., 2007), some of which have an infrastructural character. In this section, we zoom in on another aspect of what is needed to put data to work. Part of this work is facilitated by the use of conventions and standards. These help to organise data, to collect and handle metadata deemed necessary to make sense of that data, and they make it easier to combine and compare data. In our discussion of geographical data, we noted that the GPS system connects to a reference system, the World Geodetic System (WGS 84), which links position with location. A convention such as this connection to a reference system makes it possible to layer data. Very concretely, the possibility to switch from map to satellite view in Google Maps depends on the conventions about which reference system to use and how to use it. Such conventions are used for all data types, to make scientific research possible with brain scans (Beaulieu, 2002) or for entertainment purposes, like playing Pokemon GO.

Conventions are labels that cover many types of agreements. Conventions can take the form of a protocol (an agreement on process, which steps to follow) or standards (an agreement on the measure of quality, values or format). They can be about data formats, file systems, metadata or about using object identifiers, in which case they are important for automation and for large-scale data collection. Conventions can also vary in their degree of formality. They can be very detailed and internationally implemented, for example, an International Organization for Standardization (ISO) mandated standard. They can also be very informal and local, for example, how to annotate data in a spreadsheet used by a few colleagues who collaborate intensively. All these conventions make it easier to link data across different datasets and to give confidence in the combination of data from different sources.

While this may sound very technical and bureaucratic, conventions also have an important role in shaping the value of data and how confident we are in using them. If there is a strong set of conventions around a type of data, and if these are implemented in similar ways across different locations where data are produced, then we are more confident that the data are comparable and can be sensibly aggregated. An example of this is drug-testing protocols: we want to be sure that the data are produced and handled in the same way across the different test locations, so that we can aggregate the data and have a sense of whether the drugs have a positive effect. This is especially important in a context where we want to avoid unnecessary risks because human health is involved, and where we want independent confirmation of effectiveness because a lot of money is at stake for the companies that claim to produce effective drugs. Conventions help us navigate such situations. They contribute to 'quality' and trust, because they assist in maintaining data integrity, establishing provenance and preserving privacy. They can also help us discover whether data have been modified, or whether data have been removed from datasets and might lead to a skewed view, for example.

Metadata is a specific category of conventions. Metadata is structured data that describe datasets or documents. It helps to make sense of their contents, of how they might be related and of their history. Metadata is therefore a type of notation that helps understand how data are structured and that makes it easier to use or manage data. Metadata can be about who is the owner of the data, when and how they were collected and by which means. An everyday example is a barcode. Another example of metadata is a DOI, a digital object identifier, which is linked to a publication. When you use an application like Zotero to save a reference from a webpage, Zotero interacts with the metadata of the webpage to extract the bibliographic reference and import it into the Zotero database. As in the case of data, whether something is metadata depends on the use made of it. A library coding system can be considered metadata, but for a historian writing a history of classification systems, the coding system would be data that are informative about the kinds of classifications; for example, when did 'young adult' literature as a separate category become common? Metadata can also change over time (Li and Sugimoto, 2017).

As data circulate and data-sharing intensifies, it can be critical to keep track of where data originated and how they were produced. The use of metadata to document provenance is a common strategy, not only in biomedical data-bases (Van Horn and Toga, 2009), but also in data-sharing platforms across life and social sciences (Dormans and Kok, 2010) and in social and cultural pro-duction (Beaulieu et al., 2013; De Rijcke and Beaulieu, 2014). Scientific meta-data helps to create common ground between different users in different institutions or disciplines (Edwards et al., 2011). There are of course many ways of describing a dataset, and to order these descriptions, 'metadata sche-mas', 'ontologies' or other types of semantic tools are used. Metadata schemes are often developed for specific types of objects. For digital documents, a widely used schema is the Dublin Core. For websites, the Semantic Web aims to formalise what we know about their contents. For research data, a useful standard has been set by the OBO Foundry, which provides principles around which various computational ontologies pertaining to different scientific domains can be built. As we discussed in Chapter 3 with regard to call data records, much metadata about phone calls can be very useful to find out about people, even if the content of the conversation (what you could arguably call the data) is not known.

In Big Data settings, metadata is often referred to as annotations or descrip-tors. When data from different sources and in different formats are stored together in a repository without a predefined schema (as a way to avoid data silos, which occur when you store data in data warehouses), metadata is used as a way to structure data. In such a 'data lake', data remain usable thanks to metadata, since it makes it possible to query data. Again, dealing with the size of metadata is a challenge, since it must remain accessible to users when data are accessed. While we might think of metadata as 'labelling' of data, it is also an essential layer of any information system and it plays an important role in whether it is even possible to use data. When appropriate and accessible to human or machine processing, metadata enables interaction between objects, such as data, and activities such as discovery, retrieval, provenance tracking or calculation (Greenberg, 2017).

If we think back to the GIS systems discussed earlier, we now see that metadata is needed to make the data from these systems usable. Data about location can only be used reliably in conjunction with metadata about the satellite orbit parameters and metadata about the spatial resolution available. The central role of metadata points to further issues. Because much of spatial data is used 'in real time', it is critical to have rapid retrieval and access of both data and metadata – complex puzzles for the efficient handling of data. When some transformations of data are black-boxed, it is not possible to document those transformations in the form of metadata, which also affects the accountability and therefore trust we have in data. Handling and producing metadata are therefore important issues that are entwined with data quality, computing challenges and even the use of algorithms.

5.5 Models

In Chapter 4, we hinted at the central role played by models in informing the use and interpretation of data. A data model is the result of the effort made to structure data, so that they can be analysed (using statistics for example). A data model is often represented in graphical form (see Figure 5.1). A data model is a necessary step in data analysis: it involves making decisions about how to order and visualise data. This step teaches us about what is being modelled, because the data are selected and ordered in order to represent a specific phenomenon (Leonelli, 2019). In Figure 5.1, for instance, the data model shows which categories and features are considered important for human resources processes. There is room in this data model for elements like education and salary, but hobbies or religion are not part of the data model – they are not considered relevant for what we are modelling. The data model has thus identified and restricted the part of reality that data can be used to document.

This is a fairly simple example, but the same logic applies for different types of data being modelled. The data model makes some aspects of the data count and not others. In deciding what counts as usable data, researchers define what data can matter and how. The decisions made in setting up the data model shape the range of phenomena that they will be able to consider once they start clustering and ordering data in ways that may help to interpret them as evidence.

Data models help us to capture specific aspects of the world and to pursue analyses. If we do not have hobbies or religion as part of our data model about human resources processes, we will not be able to include those aspects in our claims about efficiency of recruitment or about how careers develop. That is the reason why data modelling involves explicit discussions of the value of data as evidence for making representational claims. Data models are where evidential and representational considerations meet. When developing data

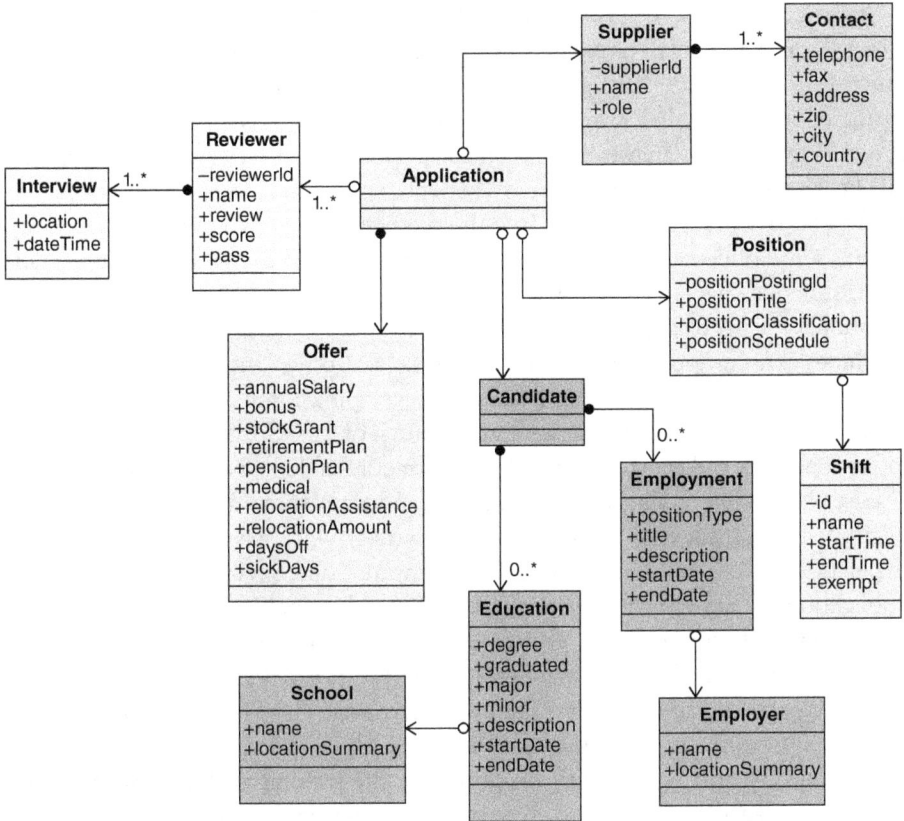

Figure 5.1 A data model for data on hiring in a private company

Source: https://www.itpedia.nl/2011/01/17/organisatie-rond-modelbeheer/. Reproduced with kind permission of Wim Hoogenraad.

models, analysts consider – implicitly or explicitly – how ordering the data will make them usable as evidence.

A second kind of model is very important if we want to put data to work: **statistical models**. While data models are usually expressed in graphical form, statistical models are expressed using mathematical notation. Statistical models enable us to describe the data in ways that make them manageable: they are precise and concise descriptions that enable calculations. Statistical models help us make claims because they encompass important aspects of the relationship between variables, and between the data we collected and the world. They guide the methods used to extrapolate patterns from data. They help us evaluate whether or not such patterns are meaningful, and what 'meaning' may involve in the first place. Statistical models are therefore central to the credibility of the data-driven approach.

Statistics can help us ask questions like do the data really show an important effect or do data really correspond to a (consistent) phenomenon in the world. In other words, we use statistics to explore the validity and reliability of patterns extracted from data. For instance, statistics is often hailed as a powerful means to detect error within datasets in relation to specific hypotheses (Mayo and Spanos, 2011). In a Big Data context, there is a set of statistical models and techniques that tend to be used most frequently (see the discussion of correlation in Chapter 2). Hence, some philosophers and data scholars have argued that 'the most important and distinctive characteristic of Big Data [is] its use of statistical methods and computational means of analysis' (Symons and Alvarado, 2016). Examples of this are machine learning tools, deep neural networks and other 'intelligent' practices of data handling. There is a lot of emphasis on developing computational methods for data analysis within Big Data research, and much of the efforts focus on improving inferential tools and methods, in order to extract reliable knowledge from data.

Last but not least, computational models and algorithms are also a central type of model to consider for understanding how data work is structured. Statistical expertise needs to be complemented by computational savvy in the training and application of algorithms associated with AI. This includes machine learning and other mathematical procedures for operating upon data (Bringsjord and Govindarajulu, 2018). These are closely linked to the statistical models selected.

Consider for instance the problem of overfitting. This is the mistaken identification of patterns in a dataset, and the result of imposing a model on the data too rigidly. Overfitting is greatly amplified by the training techniques employed by machine learning algorithms. There is no guarantee that an algorithm trained to successfully extrapolate patterns from a given dataset will be as successful when applied to other data. It is possible to minimise this problem by reordering and partitioning both data and training methods. This makes it possible to compare the application of the same algorithms to different subsets of the data. This is called 'cross-validation'. Another approach is to combine predictions arising from differently trained algorithms ('ensembling'). A third technique is the use of hyperparameters, which are used to constrain the learning process. To do this well requires knowledge of the mathematical operations and of their implementation in code, as well as familiarity with the hardware architecture (Lowrie, 2017). In other words, working with models from statistics and mathematics needs to be complemented by expertise in programming and computer engineering (the need for this layered knowledge is discussed in the next chapter).

The point is that structuring of data work in data science is different from data analysis in statistics, as developed over the past century in the social and natural sciences. Whereas regressing or rule-based deduction is used in traditional statistics, machine learning builds programs that develop their own approach to data description (Lowrie, 2017). Focusing specifically on computational systems, John Symons and Jack Horner (2014) argued that much of Big Data research consists of software-intensive science rather than data-driven research. These elements structure how we can put data to work: the production of knowledge claims depends on the manipulation of models implemented in database design (data models), in analysis (statistical models), and within software and computation (machine learning and algorithms).

5.6 Visualisations: Forms, tools and interfaces

5.6.1 Data visualisations

Data visualisations are a way of ordering, encountering and interacting with data. They differ from the graphical data models discussed above (Figure 5.1) insofar as they are aimed at conveying the data rather than the properties of the data. As such, data visualisations are considered to be of much wider interest than the more specialised data models that are mainly used by data workers who deal with databases and computational work. Data visualisations circulate widely and shape what we know and the questions we ask of data. Visualisations are themselves shaped by data practices and technologies. Data visualisations are often presented and perceived as the way of letting data speak for themselves, but they are neither transparent nor self-evident. Visualisations are 'acts of interpretation masquerading as presentation' (Drucker, 2014, p.16). In this section, we consider how visualisations have developed over time, the main conventions that shape how data are visualised, and how data visualisations already contain selections and interpretation of data.

We discussed earlier how data are not merely representations of phenomena. We extend this argument to data visualisations: these are not merely representations of data. Data visualisations are the result of many steps, and our appreciation of them as visual renditions depend on conventions that are often so familiar that we don't notice them. What distinguishes a data visualisation from other types of expression (text, numbers) is that they use space in a meaningful way. That is to say, the conventions for displaying data *spatially* are a central component to making data visualisations meaningful. These conventions are often forgotten, but they are both fascinating to understand and important in order to learn from visualisations.

Data visualisations can be organised according to their:

- graphical format (map, table, chart, network diagram);
- purpose or function (navigating, record keeping, calculation);
- type of content (spatial, qualitative, quantitative, temporal, interpretative);
- the way they structure meaning (analogy, connection, comparison, multi-variate, axes);
- disciplinary origins (bar diagrams from statistics, trees from genealogy, flow charts from electrical circuits) (Drucker, 2014).

Across the variations in the types of data visualisations, all have a number of processes in common (Drucker, 2014). The first is the rationalisation of a surface. This is the process of setting a space apart so that it can be meaningful; for example, the separation of the space on a page between the space for the running text and the space for a figure. The second is the distinction between figure and ground. Creating a contrast between an object and the background directs out attention and tells us what to pay attention to. A third process is to have the visual elements of a figure work together, in a relational system. This could be the framing of a visualisation or the use of a legend that enables us to put the visual element in relation to a shared reference. Together these processes create what we consider to be a visualisation. Within a visualisation, the organised space can then be used to express meaning. Spatial relations like proximity, hierarchy and juxtaposition indicate how to understand data (Drucker, 2014).

All these visual conventions contribute to make data visualisations work in specific ways. They tend to be treated as transparent (Kennedy et al., 2016), whereas they are already carrying some interpretation of the data. An illustration of this is the reference systems we discussed with regard to GPS data. When we look at maps of the world, we are aware that a three-dimensional space has been brought into a flat, two-dimensional surface. This is done via projections, and different projections take different elements into account, but we pay little attention to this convention because we are so used to it.

Naturalised maps and charts come across as 'what is' – as being the same as the phenomena. When we stop paying attention to which assumptions are taken into account, we start to see data visualisations as transparent. This applies to all visualisations, even something as simple as a bar chart: we tend to forget all the work that goes into producing these visualisations (sampling, smoothing, colour selection) and how they produce meaning (Drucker, 2014). In Figure 5.2 an association between maleness, genius and scientific ability is expressed in the use of similar shapes (squares). Such associations matter because visualisations shape how we experience the world. According to this chart, females could pass on these traits, but never express them. By visually reinforcing that the symbol for men is related to the symbol for genius, the visualisation reinforces a sexist interpretation of the data.

Data visualisations are closely associated with Big Data and are a common component of data science training. That visual information constitutes the richest mode of input to human cognitive systems (so-called 'power of the human visual system') is an assumption that has a long history, going back to at least the middle of the 20th century. Each visualisation contains assumptions and principles of knowledge, and a particular point of view. Many conventions from older modes of observations persist, even when we are dealing with large-scale information that is beyond human perception. Rather than windows that give us a perfect view onto data, visualisations are better understood as lenses that are imperfect, but useful precisely because they are selective (Beaulieu, 2001).

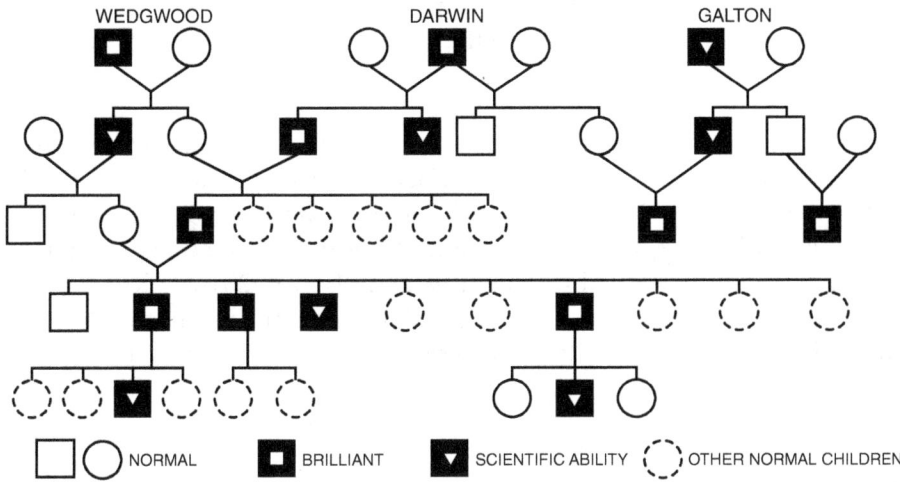

Figure 5.2 Conventions of visualisation shape the contents: the icon of genius is visually related to the icon of maleness

Source: Chart produced by Francis Galton, for the English Eugenics Society. Chart showing inheritance of ability, Eugenics Education Society. This poster was commissioned by the English Eugenics Education Society and designed and produced by Philip Benson's London advertising agency in 1926. Reproduced by kind permission of the Museum of London.

5.6.2 Infographics

In the past decade, the format of the infographic has become a prominent way to communicate visually with data. The term infographic is a contraction of 'information graphics'. Infographics vary in style, but all have in common that they are visual representations based on statistical data. Their history is sometimes taken to go back to the work of Florence Nightingale in the 19th century, but there has been an explosive increase in their usage since about 2010, related to the availability of web-based tools to create and circulate infographics. Typical of infographics is the combination of graphical and other elements, assemblage of text and numbers, charts, graphs or maps, and characters to transform data into visually accessible arguments (Featherstone, 2014).

Infographics can be:

- statistical graphs (visualisations that rely on quantitative data without adding another layer);
- data maps (combining cartographic form, variables of distance and visualised quantitative data);

- time series (which place quantitative data within a visual temporal context);
- relational graphics (which utilise composition to convey differences in size between variables and analogise the properties of datasets with the physical world) (Tufte, 1983).

Across all these categories, it's important to note that digital infographics are sites of intersections between interfaces, people and data (Amit-Danhi and Shifman, 2018). Furthermore, in a digital context, infographics are not only to be looked at and understood, but like other visualisations, infographics can act as interfaces and sustain interaction (Frosh, 2016). They can also give a more layered sense of engaging with data (Amit-Danhi and Shifman, 2018). Like other kinds of data visualisation, they are interpretations.

5.6.3 Data visualisations as interfaces

One of the important features of recent forms of visualisation is that they merge the function of engaging with data visually with other forms of interaction. Digital visualisations have tended to merge with the artefact of the graphical user interface. This means that seeing and doing are brought together. These 'visualisations as interfaces' invite the user to zoom in, select, or otherwise interact with data (de Rijcke and Beaulieu, 2011).

As users of interfaces different actions are open to us. Depending on how the data resource and interface are built, we can view, filter, configure, select or construct data objects. Some of these interactions are quite shallow, while others allow deeper interaction with data. Infographics tend to be at the shallow end, and the interactions they allow are quite limited. What is important is the way these interactive possibilities 'create a sense of data experience' (Amit-Dahni and Shifman, 2018). By engaging with the data themselves, the viewer/user touches the evidence and is put in the position of seeming to have direct access, to be able to explore and draw unbiased conclusions. The viewer/user can make the data speak for themselves, but only within the set of infrastructures, conventions and rules of visualisation we have discussed so far.

A final aspect of data visualisation to note is the increased personalisation of data visualisation. Across many apps, it is now very common to have access to data visualisations that can seem like intimate self-portraits. We see ourselves through detailed data visualisations that show how our bodies and activities can be mapped out in time and space. A daily consultation of our mundane practices has become common practice: we are used to seeing graphs of our sleep, our movements, our hoped-for path home... and to responding to cues from such visualisations for our sense of self (you run faster than 86% of users) or to script our lives (move more, drink more, stress less).

Through these visualisations, the very relationship between objects of knowledge and knowing subjects is changing. New interfaces with data are created, such as (dynamic) visualisations. These not only aim to 'show' data, but also emphasise their dynamism through live feeds, interactivity and the possibilities to layer types of data. Visualisations also function as interfaces to data, to explore and act on them. Visualisation techniques make data seem 'transparent and accessible' while the underlying models are opaque (Thayyil, 2018). As we have seen however, visualisations are far from transparent. Critical skills and awareness of the kinds of data assemblages that enable these visualisations are indispensable to using data visualisations responsibly.

5.7 Curation

The final element in putting data to work that we will discuss in this chapter is curation. Curation – or, as it is also sometimes called, 'data management' – is perhaps the most undervalued aspect of data work. Yet, it is necessary for successful data circulation and use. Data curation is the process of organising and integrating data from different datasets. It is guided by the desire to maintain the value of the data across the adjustments needed to make them usable. As such, data curation requires a strong understanding of both the data's creation and their potential further uses.

To illustrate how significant the work of curation can be, think of the experience of grocery shopping in a different country. In the United States and Canada, sugar is located in the baking goods section, together with flour and baking powder. In The Netherlands, sugar is located next to the coffee and tea. Sugar is therefore categorised very differently in these two contexts. In the case of a supermarket, it's possible to walk around until you have found the desired item – for a database, the labels assigned to data are the only way to navigate the database. To extend the analogy, how could you ever find the sugar if you look for it with a label of 'baked goods' in a Dutch database? Data curation is therefore fundamental to the usability and quality of a database.

As global data infrastructures, interfaces and related institutions become ever more sophisticated, the resources needed to curate them have also grown exponentially. Within the private sector, the increasing costs of storing and analysing data have heightened their value as commodities, with increasing efforts to license datasets so that they can be incorporated into specific property regimes (more on this in Chapter 8). Companies are also frequently looking to either outsource data maintenance (i.e. by using external web and cloud services) or forage for easily accessible data (as in the case of social behaviours expressed in publicly available social media). The extent to which data are circulated, and the restrictions under which such circulation can be placed, can thus vary dramatically.

Within academic research, all the work required to put data to work does not fit contemporary regimes of funding, credit and communication. Monitoring data infrastructures and keeping them up to date requires serious investment, without which the quality and reliability of the Big Data used to fuel AI tools cannot be guaranteed. As we will see in Chapter 7, the more data move around and are used for a variety of purposes, the more vulnerable they are to unwarranted and even misleading forms of manipulation and enrichment. And yet, funding agencies focus the vast majority of their resources on supporting novel research and rewarding the publication of high-impact scientific papers. Long-running infrastructures do not fit this model of evaluation, since their core business is conservation rather than innovation (though of course they do require constant updates and the uptake of new technologies).

Data curation is done by a variety of experts whose training and titles can vary. They include librarians, information scientists, project managers, consultants and research assistants. Their work can address both upstream and downstream management of data, from the point of data creation to the archiving and sharing via repositories (Palmer et al., 2017). The creation and evaluation of metadata are part of data curation. While we noted that the production of metadata was sometimes automated, there are many aspects of data curation that cannot be formalised and require a high level of familiarity with data creation, competence in overseeing infrastructural aspects and expertise on the needs of users. Even when automatically produced, informal communication about metadata also helps to make data more useful (Edwards et al., 2011). Curation therefore plays an important role in shaping or packaging data and making them intelligible to users who were not part of their creation. This work often seems technical, outside the more valued activities of doing research. Yet the use of labels to describe data (sometimes called ontologies) is vital for users to be able to retrieve the information they need from a database. A shared understanding of what these labels should be (as in the sugar example above) is very important.

The creation of metadata 'by hand', through annotation and filling in of information by humans for each data point, is highly time consuming. In some fields, there is a high level of professionalisation around metadata, but in many fields this work is considered part of the researchers' tasks. Creating metadata can feel like a burden on top of a scientist's primary work. As Edwards et al. (2011) explain, 'Research scientists' main interest, after all, is in using data, not in describing them for the benefit of invisible, unknown future users, to whom they are not accountable and from whom they receive little if any benefit.'

Those who have the expertise to maintain and curate databases are often overlooked and undervalued, since they do not routinely publish in top-ranking journals and may therefore not be recognised or rewarded as high-level researchers. This affects both the status and the availability of data curation positions at research-performing institutions. It is difficult to make a case for creating jobs in data curation, and even when they do exist, they are often ranked as 'service' jobs (on a par with technicians) rather than as 'research' jobs (and thus seen to contribute directly to knowledge creation), with serious consequences for the career prospects and salary scales of this type of data worker. There are therefore few incentives to enter this path of work, even though it is central to the large-scale mobilisation and reuse of data that power contemporary science.

The care for data does not fit the current rhetoric around Big Data. It requires work and time, it is not exciting and it is expensive. This creates the risk that data are badly managed, unreliable, unfit for repurposing – and that because they have not been curated well (e.g. without significant metadata), they cannot be recontextualised adequately.

5.8 Conclusion: Forms of data work

In this chapter, we have gone beyond data journeys to explore how data are put to work. We have looked at what is needed to circulate, share or reuse data. We considered putting data to work from different angles and, for the sake of clarity, discussed in turn the technological, infrastructural, communicative and labour dimensions of putting data to work. Across this discussion, we constantly pointed to how each dimension relies on and affects the others in practice. There is a great variation in the extent to which these elements are formally organised, from the highly regimented databases of the WHO to the looser mash ups of start-ups using social media data. In spite of this variation, we saw that to make data work takes work. More data is not enough! The practices that enable the flow of data involve infrastructures (networks and platforms), conventions (standards, annotations), models and visualisation tools, and related expertise (curation). Together, they play a decisive role in the multiplication of the uses and users of data.

ADDITIONAL READING

Acker, A. (2018). *Data Craft: The Manipulation of Social Media Metadata*. New York: Data & Society Research Institute.

Drucker, J. (2014). *Graphesis: Visual Forms of Knowledge Production*. Cambridge, MA: Harvard University Press.

Frigg, R. and Hartmann, S. (2020). Models in Science. *The Stanford Encyclopedia of Philosophy* (Spring 2020 Edition), Edward N. Zalta (ed.) https://plato.stanford.edu/archives/spr2020/entries/models-science/.

Starosielski, N. (2015). *The Undersea Network*. Durham, NC: Duke University Press.

6

NEW DATA SKILLS

---------- Overview of chapter ----------

Summary

In the context of data circulation, the level of 'project' is central. In this chapter, we consider different kinds of data work done in projects, from the perspective of people who see themselves and are seen by others as being data workers. One label among many for these people is data scientist. In order to understand what data scientists do, we first have to discuss what is usually understood when using the term data science. It is important to consider what data work involves to really grasp how different aspects of data work influence each other and to value them. We will then zoom in on the skills that are needed for data scientists and data workers and on how collaboration can take place across these skillsets. By discussing data work in detail, we help map out a complex field and give a conceptual basis to understand why collaboration is so important to achieving data projects successfully.

6.1 Introduction: Data expertise

Imagine a project that aims to understand traffic patterns in a city. What would be needed, in terms of data work, to pursue such a project? We would need to think about which data is already available and whether additional data needs to be collected. A good plan for data collection would need to be developed, to ensure that data of sufficient quality and scope is collected. Given that the data is generated by citizens' activities in their daily life, we would also need to consider how to engage citizens and how to ensure that this data collection would not be harmful to them as individuals or as groups. We would also need to figure out how to move the data from the points of collection, to storage facilities and to where the data is going to be analysed. Such data might be especially amenable to being mapped out geographically, and we would need to figure out how best to show traffic flows and bottlenecks in the space of the city. And we would need to decide on who will be able to access the data, in which form and for what purpose. A serious set of tasks for any project team!

Clearly, doing data science requires many kinds of work and different types of expertise. This expertise needs to be coordinated. This is true not only across tasks, but also to perform a single task – to decide how much data needs to be collected, you have to understand the implications for both the computational needs of the project and for how you even conceive of traffic density. The requirement for different types of expertise to work closely together is also increasing. For example, as datasets become more diverse, it becomes even more important to link expertise on the domain in which data was created to understand the meaning of the data, with computational expertise to be able to handle the diversity of data.

There are very few individuals who possess skills across all these areas. Furthermore, projects are usually too large for single individuals to take them on. This means that expertise needs to be distributed across individuals within teams, or even across teams in an organisation. These experts must not only be competent in their own area, they must also be able to work together if they are to pursue data science successfully. In this chapter, we aim to provide you with greater awareness of the various types of expertise and skills needed. We also describe how experts can work together, by reviewing different models of collaboration. This material will enable you to better understand your own expertise and skills, to value the expertise of others and to find effective ways to combine them to work together.

6.2 What is data science?

6.2.1 A growing field

The increasing attention to data as valuable outputs in and of themselves is occasioning a shift in the division of labour within research and development. The very idea that **data science** can be a separate research domain is relatively new, and what data science actually consists of continues to be a matter of debate. This has concrete implications: there are very few people who have been trained as data scientists. Rather than having followed a programme with the title 'data science' in college or university, they have come from different disciplines, such as statistics of computer science, and, over time, have come to occupy a position of data scientist in their organisation. This diversity matters because it means that data scientists in different organisations are actually doing very different jobs and might be in very distinct departments within an organisation; for example, in marketing, business intelligence or R&D. It also means that their expertise might contrast as well, depending on their initial training and the context in which they have worked since studying. This diversity has consequences for careers: in some universities, data workers are considered professional staff, whereas in others they are considered scientific staff. All this variation makes it difficult for the professional status of data workers to be recognised and limits career development or even job security. This is an odd situation, especially given that demand for data scientists far outstrips availability, a trend that has been observed since at least 2008 (Swan and Brown, 2008).

When there is a clearly and widely recognised institutional embedding for a kind of expert, this adds to the legitimacy and recognition of a particular area of expertise. One way to establish the status of a type of expertise is to link it to a scientific discipline. Is it a question of time until data science becomes a coherent, recognised discipline, as a basis for a well-defined profession? A discipline has:

- a shared object of study and methods;
- an accepted body of knowledge;
- a community of scholars who primarily identify with the discipline;
- mechanisms for communication (publishing) and reproduction (teaching).

Whether a discipline is forming can often be traced by looking at mechanisms of communication and reproduction. In other words, whether there is a particular body of knowledge and a discourse in which a community is involved, and whether there are training programmes built around a recognisable set of core elements. With the rise of dedicated journals and the existence of hundreds of accredited training programmes, two key markers of disciplines – mechanisms for communication (publishing) and reproduction (teaching) – seem to be in place. With regard to journals, we can identify a number of publications that are well regarded and have data science as their core subject: for instance, the *Data Science Journal*, the *Harvard Data Science Review*, *Data*, the *International Journal of Data Science and Analytics* and *Patterns* (among others). With regard to educational programmes, data science degrees can be found from college to PhD level, and there are now guidelines and recommendations on what a data science degree should cover (De Veaux et al., 2017; Mikroyannidis et al., 2018). In 2020, over 250 undergraduate programmes could be found at 'bricks and mortar' universities on all continents, and more degrees were offered as online courses.

Obtaining a degree in data science is not the end of the story, however. Because of rapid changes in the kinds of data and in computational techniques, data scientists have to be willing to keep learning throughout their careers. In addition, data scientists need to be aware of the domain in which data are created – data from biology and from marketing need different knowledge to be used properly. This dynamism requires that data scientists be willing to be lifelong learners, and requires that there be support for this learning (e.g. Mikroyannidis et al., 2017).

Data science should also be understood in the particular socioeconomic and cultural context of the first two decades of this century. In the same period that data science has become increasingly prominent as a field or discipline, there have been important changes in universities. Universities were established around specific disciplines, and from the end of the 19th century in the Global North, faculties and departments became increasingly specialised. This focus on single disciplines has changed over the past few decades. For one thing, interdisciplinary education has been increasingly valued. Many programmes in data science are indeed taught via university-wide coalitions between different faculties. These tend to be the most successful, although these initiatives encounter additional administrative overheads to deal with cross-organisational entities (Berman et al., 2018). Teaching of data science is also linked to innovations in universities, in terms of how education is organised:

> Data science is by definition interdisciplinary and requires students to interact widely across academic disciplines and with non-academic partners, since they too are making rapid progress in the field of data science. This requires a new type of education that is future-proof with respect to data science. To achieve this, we also have to adapt the education system, which needs to change from the more classical way of providing education aligned with the traditional academic disciplines (Wijmenga, 2019).

Such new contours for how universities can function as knowledge institutions have been emerging in the past decades. First, there has been growth in trans-disciplinary research involving actors outside the university, often in public–private partnership (e.g. businesses, patient groups, non-governmental organisations (NGOs)). For data science, this means that students are often involved in projects developed in partnerships with non-academic actors, and that they are learning by engaging with 'real-world' problems and datasets. Second, activities of 'valorisation' have been increasingly stressed. This means not only that 'scientific discoveries' are stimulated and rewarded in universities, but that innovations, patents, start-ups and other kinds of contributions for which a 'context of application' or societal value seems obvious are rewarded. In the context of datafication of society, we can see how data science would be an area that would benefit from, and contribute to, a version of the university as a place that produces useful, economically and socially relevant or applicable knowledge.

A final important dynamic is the growing diversity of sites of knowledge production. Research is no longer primarily associated with universities, and knowledge has become central to a great many organisations (Gibbons, 1994; Nowotny et al., 2001; Strathern, 2005; Wouters et al., 2013). The results of all these changes are multiple and layered, and also vary according to funding schemes in place in different national systems of education. Generally, though, we can speak of pressures on universities to show that they are 'productive' and useful to society. In the current context, research and teaching priorities are not only shaped by criteria internal to universities, but also are increasingly responsive to societal challenges and corporate interests.

What does this mean for data science? The current trends in universities that prioritise the connection of science to (technological) innovation and economic value, and openness to the world outside academia may work against the internally focused dynamics necessary for discipline formation. In other words, data science may not solidify into a recognised discipline based in universities, in the way that molecular biology or women's studies have become recognised departments in universities. In addition, there is a strong pull on experts towards industry and away from academic institutions, to the point that it prevents institutions from developing data science programmes (Berman et al., 2018). Higher salaries and short-term benefits are not the only reason for this brain drain. The recent report of the National Science Foundation (NSF) Working Group on data science noted that an emerging problem with maintaining and developing scientific research in data science also has to do with the fact that 'when the best infrastructure environment for cutting-edge research is consistently in the private sector, the opportunity for innovation in the public sector deteriorates' (Berman et al., 2018). This brief discussion of disciplines and of universities as institutions helps understand the context in which data science is developing and why we cannot consider it solely as an academic project: corporate and societal actors are also shaping the contours of data science, and vice-versa.

6.2.2 A composite field

If it is difficult to describe data science as an academic discipline, how should we talk about it? Describing data science as field at the intersection of different disciplines or areas of knowledge is a common approach. Venn diagrams are often used to show data science as overlapping areas of knowledge, giving a strong sense of data science as a composite field. One of the early descriptions of data science can be found in Figure 6.1.

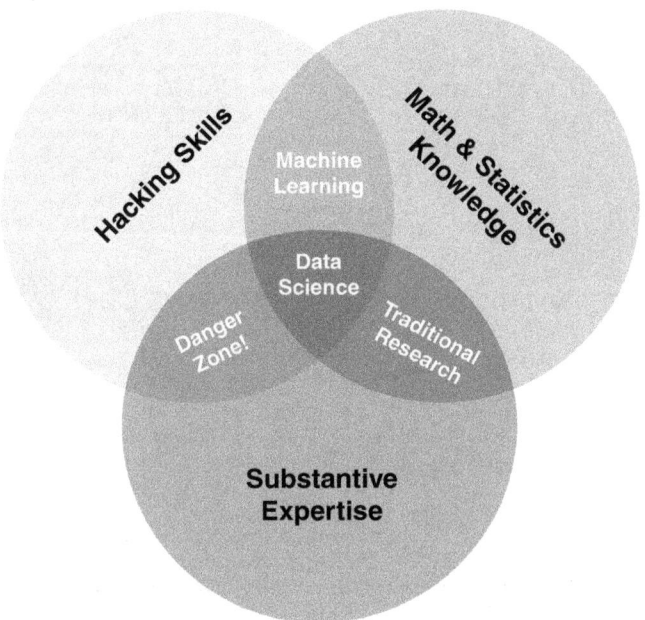

Figure 6.1   Data science as composite

Source: First presented during a talk in 2010 and later published on a blog in 2015 (Conway, 2015).

This way of representing data science seemed to resonate strongly with the data science community, and this Venn diagram led to countless variations on this theme (see Figure 6.2). The variations are interesting in and of themselves, but the main message is that data science is the result of combinations of different areas of expertise.

This one replaced 'hacking skills' – some found this term objectionable because of its criminal connotations – with computer science and specified the resulting overlapping areas in more helpful terms.

The types of expertise that need to be combined are also debated. Increasingly, there is recognition that efficient and appropriate use of data is not solely the result of combining mathematics with computer science and statistics. As Xiao-Li Meng stated in his inaugural editorial in the *Harvard Data Science Review* in 2019, data science is 'not just machine learning or just statistics' and 'not all about prediction'. Data science is not even 'only about

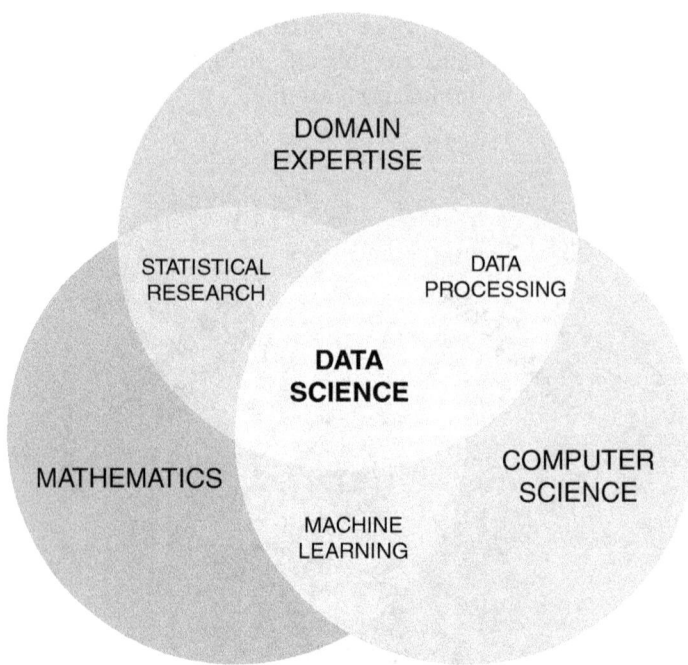

Figure 6.2 Another version of data science as a Venn diagram

Source: By Shelly Palmer: https://www.shellypalmer.com/shelly-palmer-bio/. Reproduced by kind permission of Shelly Palmer. From *Data Science for the C-Suite* (2015), and https://www.shellypalmer.com/data-science/.

data analysis' (Meng, 2019), given the many stages in the data journeys necessary to make data actually usable. Data science must also address how it gleans knowledge from the world and produces data as evidence to support claims. This means that epistemology is also a core concern that affects both the daily work of data scientists and shapes the place of data science in society. The very aims of balancing appropriate assumptions with computationally efficient approaches and other trade-offs are epistemic ones. Furthermore, at the heart of technical questions around how to integrate different data formats or how to engage participants in data collection are critical issues about data governance.

We see data science as dealing with four basic types of expertise: computational and statistical, both of which are typically regarded as core 'technical' expertise; epistemological expertise, including an understanding of where data fit in the processes of knowledge production and how different stages of a data journey may affect each other; and expertise on data governance, which

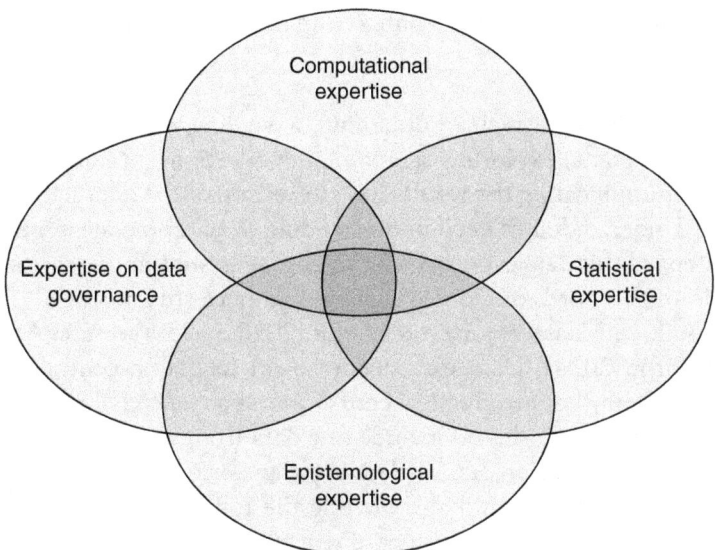

Figure 6.3 Data work as conceptualised at one of the author's workplaces

Source: Data Research Centre, University of Groningen, The Netherlands.

encompasses the usability, regulation, ethics and curation of data (see Figure 6.3). The relevant areas of expertise may change as data science transforms in response to internal and external factors. It remains, however, that while these Venn diagrams powerfully make the case for the need for a diversity of experts to pursue data science, what really sets data science apart is the need for integration across these different spheres. It is not enough to put these (or other) types of expertise next to each other and expect that success will ensue. These areas of expertise must truly intersect – rather than co-exist. This is what is distinctive about being a data scientist versus being a statistician, computer programmer or legal expert. In that sense, the proliferation of these Venn diagrams may be explained by the fact that they foreground the kinds of expertise needed, while missing out on the interactions that, we argue, are what constitute data science. For this reason, we move from a static view on the disciplinary areas that contribute to data science, to a discussion of the skills that make the dynamic process of data science possible.

6.3 Data science skills

Why should we emphasise the dynamism and interrelated aspects of data science? Consider this passage on developing curriculum for data science, and note how it links processes, context and experiential knowledge in learning data science:

> The recursive data cycle of obtaining, wrangling, curating, managing and processing data, exploring data, defining questions, performing analyses, and communicating the results lies at the core of the data science experience. Undergraduates need understanding of, and practice in performing, all steps of this data cycle in order to engage in substantive research questions. In the words of Google's Diane Lambert, students need the ability to 'think with data' (Horton & Hardin 2015, p. 259; see also ASA 2014a and Shron 2014). Data experiences need to play a central role in all courses from the introductory course to the advanced elective/capstone. These experiences should include raw data from a variety of sources and should involve the process of cleaning, transforming, and structuring data for analysis. They should also include the topic of data provenance and how it informs the conclusions one can draw from data. Data science is necessarily highly experiential; it is a practiced art and a developed skill. Students of data science must encounter frequent project-based, real-world applications with real data to complement the foundational algorithms and models (De Veaux et al., 2017).

Similarly, the US NSF Working Group on the emergence of data science stressed the view of data science as a process made up of different practices and skills that could be strengthened by building better connections across the data life cycle to reduce the current gaps (Berman et al., 2018).

Doing data science is therefore about connecting the use of data for prediction, exploration, understanding and intervention. It requires a thorough understanding of probability, of the relationship between sampling/populations, of the consequences of false positives/negatives, and a firm grasp of correlation and causation – a set of skills that should also be shared widely in society given the growing importance of data science (Garber, 2019). Doing data science also means being able to judge when to simplify or approximate, to make trade-offs to optimise algorithms, and to balance speed and accuracy. Finally, doing data science requires understanding how we shape our world and our knowledge of it by making decisions about samples and populations (Blei and Smyth, 2017; Bates et al., 2020; Francois et al., 2020). Because of the societal significance of data science, it is also important to be able to evaluate the role of data agents who are explicitly or implicitly producing and sharing data, new risks and benefits associated with this, and how configurations of access to data shape power in society. In what follows, we review the types of skills involved in meeting these requirements.

6.3.1 Technical skills

Data science involves technical skills typically associated with quantitative analysis and problem- solving. Mathematics (especially optimisation and probability) and statistical inference (including sampling) and modelling remain essential, but they are not sufficient for data analysis. Equally important are computational skills such as coding, machine learning and experimental data mining. A computational approach will focus on interrogating the data in a way that an information processing machine (a computer) can handle. It will also consider whether the analysis and handling of the data will remain feasible as the dataset grows. Design and visualisation skills are also highly sought after; as we saw in Chapter 5, how data are ordered and visualised matters a great deal to their interpretation. Another set of skills has to do with database management and data curation. How to query, retrieve and handle data, as well as cleaning and structuring data are important to be able to meaningfully explore, analyse and visualise data. The combination of these different technical skills is indispensable for data science.

6.3.2 Analytic skills

Data science requires the skill to formulate productive questions; that is, questions that can be answered with particular datasets. Though lots of people have the skills to crunch the data and answer set questions, asking the right questions is an even more valuable skill (Dumit and Nafus, 2018). In dealing with data, a valuable 'habit of mind' is to develop a critical stance towards the quality and provenance of the data. This means asking questions like 'How were the data collected?', 'How are the variables defined and the constructs operationalized?' or 'Why, for what purpose, and in whose interest was the data collected in the first place?' (Finzer, 2013).

Asking how the data were collected may seem like a simple descriptive question. However, to use the data well and responsibly, we also need to understand the meaning of interactions around and through data. This is especially important because data used is increasingly created through interactions with networked and mobile data devices used by individuals in daily life. A device like a mobile phone can be an intimate possession, used to quantify the self, and to integrate and manage ongoing intimate relationships. It can also be a common possession that is casually shared between several people. In the first case, the connection of the technology with an individual self will be a valid assumption – and we can equate data from one phone with data from one individual. In the second case, it will not – data from one phone will be data about different individuals. A powerful demonstration of the variation in use and meaning of mobile digital devices is the analysis of mobile phone use in Sierra Leone by Erikson (Erikson, 2018). She shows how a mobile phone can have multiple users and how single users have multiple phones. These practices arise in interaction with markets, with infrastructure and corporate models (multiple phones can be an economical strategy when coverage is highly patchy and using another network than one's own provider is very costly) and with the way the phone is perceived as personal or as a more fluid object (Erikson, 2018).

Paying close attention to how data is collected and how it is framed is also important to maintain a critical stance towards technology use. This makes it possible to keep questioning assumptions, and to make visible the filters through which Global North actors see the relation to technology of Global South actors. This means that in studies of the Global South users, 'their new media practices are predominantly framed as instrumental and utilitarian, partly because development agendas drive this research with a strong historical bias towards socioeconomic impacts' (Arora and Rangaswamy, 2013). For instance, there is more emphasis on farmers checking crop prices online than watching pornography on their mobile devices (Rangaswamy and Arora, 2016). To be able to understand these interactions and to avoid making incorrect or universalising assumptions about the meaning and import of data are very important skills.

Besides a thorough analysis of data quality and provenance, other habits of mind are also advantageous to learning to think with data across the different steps of data work:

- Acknowledge the need for data to gain insight.
- Look for the data: Ask 'Which data could be helpful to reach conclusions, get insight or construct arguments?'
- Graph the data: Construct graphical representations that highlight potentially useful patterns in the data; patterns that are difficult to discern by staring at a table of numbers.
- Become immersed in the data: Use (and invent) measures. Look for and tell the story behind the data (Finzer, 2013).

Because data science is increasingly being used as a decision-making tool and because of the growing role of data in shaping our world, ensuring the quality and reliability of analyses and algorithms is also important. This requires the skill of making the connection between data management and analysis, particularly to enhance the reliability of algorithms and veracity of their outputs, and the accountability of knowledge acquired through data-intensive methods. Doing this well requires not only a solid understanding of statistics but also the ability to understand data journeys. For example, being able to clean a dataset requires knowledge about what algorithms are able to do well and knowledge of their ability and limitations in detecting erroneous data. In one of the labs where one of us worked, condoms filled with water were routinely used to calibrate brain scanners. This enabled technicians to test and make technical adjustments to the scanners, without the inconvenience of having to put a person in the scanner. The technicians knew that the data analysis software was unable to systematically remove these scans from datasets and that they were likely to end up being processed as data, along with scans of brains. You can well imagine how this would skew results in very odd ways! Technicians therefore regularly intervened 'manually' in this otherwise highly automated image analysis pipeline, explaining 'Computers are not nearly as good at recognising garbage as humans are'. Being able to work in this way requires an openness to really understanding the specifics of data-producing technologies, algorithms and their interaction with data. This skill can be developed by being exposed to a diversity of data practices and by maintaining a healthy understanding of the limitations of automation.

6.3.3 Contextualising skills

Ultimately, much of data science work requires decisions with ramifications that go well beyond the setting in which a given tool, code or model is being produced. When deciding on which data to use, which algorithm to develop, which problem to tackle and how, data scientists are supporting specific ways of building the digital world that can have immediate and significant effects on everyday, material life. To address the potential implications of their work, data scientists are therefore in need of contextualising skills.

To illustrate this, consider the very vehement and weighty discussions about sex/gender categories and their construction. Across datasets, gender is overwhelmingly put forth as a binary. What would it mean to reconsider these categories? There is ongoing research to consider alternative, more inclusive ways of inquiring about sex/gender in population surveys, by using expansive check-all-that-apply gender identity lists or even write-in options (filling in a free text box) that offer maximum flexibility. Another approach is to use a multi-dimensional measure to capture different dimensions of sex/gender: sex assigned at birth, current gender identity, trans status, lived gender, and hormonal and surgical status (Bauer et al., 2017). This improves the quality and applicability of data collected and can have significant effects. Consider that trans people experience large health and employment disparities. The binary sex/gender data system used in most large population studies makes this group completely invisible and therefore makes it systematically much more difficult to address inequality. Making such changes to the use of categories can improve the data to be collected. But what are the implications for using data we have already collected? How would new categories be mapped on to old ones? Could some of the variation that is currently forced into the male/female binary be retrieved from the data? Or would this be doing violence to the data, imposing an anachronistic understanding of sex/gender? Such questions matter, especially in current data systems that are very much focused on features of individuals and tend to erase social context.

To tackle these challenges, data curators, information management specialists and experts in data studies with contextualising skills are taking a more prominent place alongside other data experts. While there is still a premium attached to individuals who can deliver an ingenious technical solution or shortcut, recent concerns with the popular image of AI have occasioned more interest in the ability to think about long-term consequences and about the contextual nature of the data structures used. For example, social science research can help understand the categories through which we make sense of our social and material world and how these are embedded in data science systems and in AI. Qualitative research is explicit about data collection and about the interventions that researchers make in the world, by creating conditions for observation or data gathering. These practices can greatly help data science to situate itself in the world and in making better design and implementation decisions (Sloane and Moss, 2019). There is increasing interest in individuals who have the skills to negotiate the intricacies and implications of decisions concerning modes of data access and sharing, and related legal and financial questions around who actually owns the data, whether and under which conditions data sharing constitutes an infringement of privacy and other individual rights, and what credit, if any, the original creators of a dataset should get when others successfully reuse their data. In the corporate world, data is no longer simply input for management or logistics, but can form the core of an enterprise's business model. In this context, businesses will also want data scientists to have affinity with business models and money-making. Data scientists will be expected to find novel ways to capitalise on data that is available (we delve into the value and reuse of data in Chapter 8).

6.3.4 Communication skills

A final set of skills are those related to communication. Data scientists must be able to communicate with other team members as well as with stakeholders who may be closer or more distant to projects. In addition, communication with different publics is also important, via infographics or other visualisations. It is therefore very valuable to be able to communicate across a range of forms (oral, visual, textual) and with different groups.

As we've noted across different discussions in this book, data is heavily shaped by its context of production. Sharing data, even between academic research groups, is a complex undertaking. To get different research cultures to share information in ways that are intelligible and accommodate different views on the world and different values is a challenging task (Hilgartner, 1995; Beaulieu, 2001; Hine, 2001; Leonelli, 2016a). For example, the areas of biodiversity and ecology research would seem to share a deep concern for gathering and sharing data on where species of animals can be found. However, ecologists are very much oriented to documenting the kind of location (the particular ecosystem), whereas biodiversity researchers are concerned with the absolute location (geolocation). These differences translate 'down' to how the data fields of databases are ordered, as well as 'up' to the way the data is used as evidence for decreasing biodiversity or disappearing ecosystems. The communication skills needed to address this are very important: in such a situation, the data worker needs to make clear where the differences between understandings of a concept like location are coming from, the possibilities for implementing these differences in both the architectures and data flows of the digital systems, and to communicate the consequences of such decisions on implementation. The same holds for communicating the meaning of categories in surveys or the respective goals of different users. Furthermore, it is only in conversation with users that the data infrastructure can be designed or adapted to suit their multiple needs – again highlighting the importance of communication skills. The communication challenge only increases as users of data diversity and different kinds of industry, government agencies, citizen groups and academic institutions become involved with data. Of course, when successful, such communication across actors can be especially powerful and stable knowledge production systems can emerge across disciplinary, organisational and national borders (Edwards, 2019).

6.4 Bringing Skills Together

Let us now reconsider the thought experiment from the opening of this chapter, where we imagined what would be needed to put together a data science project on traffic patterns. Read through Table 6.1 (inspired by Finzer, 2013) and consider the tasks listed, as well as who has the required expertise to pursue these tasks and which skills would be most essential. Note that the column primary expertise refers to the types of expertise in Figure 6.3.

From this table, we see that multiple combinations of expertise and skills are needed to undertake such a project. It would be quite extraordinary for a single person to possess all the expertise and skills needed, and, on top of that, be able to deploy them in different ways in different projects. Instead, data scientists tend to have a particular combination of skills with which they have the most affinity and which they further develop. They work in teams where their

Table 6.1 Components of a data science project on traffic patterns, divided by tasks, primary expertise and main skills required

Illustration	Task	Primary expertise	Main skills required
What data do we have about traffic in the city? Which approaches might we use to collect more data?	Review existing data sets	Governance Epistemological (reflexive domain expertise)	Analytic Contextualising Technical
What else do we want to know and which evidence do we need to collect?	Decide what data should be gathered	Epistemological (reflexive domain expertise)	Analytic Communication
How can we engage enough citizens living and travelling in the area and how to ensure that the data collection will not harm or disadvantage individuals or groups?	Data collection should be designed to ensure privacy and avoid discrimination for individuals and groups; data subjects should be involved and informed	Governance Statistical	Contextualising Communication
How are we going to gather and store data from sensors in the city?	Set up a server as a repository of data streaming in real time from a large array of geographically distributed sensors	Computational	Technical Contextualising
How will we ensure that the data are flowing smoothly?	Develop data pipelines/ process to annotate, filter and clean the data	Governance	Technical Communication
How will we assess the quality of the data and remove 'impossible' data, for example, a sensor indicating that a truck that is 300 m long rode through the historic city centre?	Explain the origin of outliers in a particular dataset	Epistemological (reflexive domain expertise)	Analytic Technical
How will we assess whether the patterns in traffic are specific to the area studied?	Decide to what extent the conclusions drawn from analysis can be generalised	Statistical Epistemological (reflexive domain expertise)	Analytic Technical

Illustration	Task	Primary expertise	Main skills required
How will we present the data to other researchers, to the city's policy makers and citizens?	Design a data visualisation suitable for publication in an article	Computational	Communication Technical
Should we use health records to examine traffic and air pollution effects?	Decide whether certain disparate datasets can be meaningfully merged	Epistemological (reflexive domain expertise) Statistical Governance	Analytic Contextualising
Which factors are most important to analyse the data?	Reduce the number of variables that need to be considered for a particular analysis.	Statistical	Technical Analytic
How can we keep collecting data securely over a long period?	Set up a version management system for data that will be gathered over a number of years.	Computational	Technical Contextualising
How can this project benefit as many researchers as possible?	Ensure proper data curation, deposit in a repository and licensing	Governance	Communication Technical

strengths can be complemented by those of others with different backgrounds. It is very common within large organisations to see data science groups include people trained as statistician, a computer programmer, an information management specialist and a social scientist, for example. Larger teams can be composed of several specialists in each of these areas. In some contexts, data curators and visualisation specialists will also be involved. Simply setting people with different expertise to work on a joint project is not enough to form an effective team. What makes a successful data science team lies in the ability for these different areas of expertise to work together: the key is collaboration.

6.4.1 Importance of collaboration in data science

Why is collaboration difficult? While interdisciplinarity has been growing in universities, disciplinary training is still very strong in most educational programmes. Academic training socialises students to become members of disciplinary communities of practice (Lave and Wenger, 1991). They are taught by mentors and teachers, they develop similar experiences by doing lab work or practicals, and come to share a set of skills, ways of doing things and of pursuing work. This means that a discipline orients its members to a particular domain of enquiry, certain ways of defining problems and agreement on what counts as evidence. When we speak of turning a business question into a data science question or of reconciling the different ways of defining location as in the example of ecology versus biodiversity, we are talking about overcoming disciplinary differences. This work is sometimes described as translation or brokering. It involves language and terminology, but also ways of seeing data. When working in multidisciplinary teams, members have to learn to interact across these practices and orientations, which means being aware of one's own approach and being able to understand how it differs from that of others. Disciplinary differences can run very deep and affect not only the direction of a project but also what counts as making progress in the project. In one study of biomedical scientists who worked with data scientists, the bioresearchers used intermediate results of the project to revise their initial research question. To the biomedical scientists, this was a positive outcome since they were now asking a better question. For the data scientists whose goal was to transfer the initial research question into a well-defined data science question and to resolve it by using machine learning and optimising performance, this felt like they had wasted their time working on the initial question (Mao et al., 2019). Collaboration between experts is challenging and the extent to which it shapes a project should not be underestimated.

Collaboration also requires that all data scientists, no matter their expertise, understand the need for different profiles in a team and the role that these other skills can play in data analysis. This is one of the reasons why data science teaching and training increasingly seeks to provide students with skills for multidisciplinary teamwork. Two common approaches to helping students develop these skills are to set up interdisciplinary teaching teams, and to expose students to 'real-world' rather than textbook problems. The idea is that these experiences will help students understand how the different aspects are entwined and the variety of expertise needed.

6.4.2 Types of collaboration

As we noted above, collaboration across disciplines has been praised for being more open to messy, societal challenges, and for supporting a diversity of sites of knowledge production. This means that collaborative science might be more likely to create responsive types of knowledge that lead to innovation and socially relevant research. On a day-to-day level in data science projects, the need to collaborate is often formulated much more pragmatically, as the need to get a job done. Perfect mutual understanding is rare in data science teams, as in any other area of life. Collaboration between experts and professionals has been studied by scholars from many fields, from computer-supported collaborative work, organisational psychology or science and technology studies and anthropology. All this work has not led to a perfect recipe for collaboration. A number of patterns have been observed, however, and awareness of these patterns can be a useful tool to help set up teams and to gain insight into the kinds of interactions that are likely to occur.

To end this chapter, we propose a short overview of different models of collaboration, as a way to become aware of how relationships in teams and between teams can develop. This awareness will help in recognising patterns of interaction. Following Barry and Born (2013), we propose two paradigms of interaction. In the first paradigm, collaboration is done across disciplinary boundaries, while the boundaries of disciplines are maintained. This collaboration within this paradigm is often labelled multidisciplinary or cross-disciplinary. Within this paradigm, one model of collaboration is the integration of the outcomes of different approaches to provide a synthesis of results. You can think of this as the 'happy family' model, where families might grow through marriage, and where the family culture is enriched by new additions – individuals interact harmoniously while maintaining their differences. In a data project, imagine team members pursuing their work according to their expectations and training, and exchanging the results of their efforts with other team members. In this model, there is sharing across disciplines, but members maintain their way of working and their assumptions are not questioned.

A second type of collaboration in this paradigm (where disciplinary boundaries are maintained) is more hierarchical. The relations between disciplines are organised so that one is subordinated to the other. You can think of this model as an 'upstairs–downstairs' situation, where one discipline provides a service to the another, which is more powerful (some data scientists are more equal than others). Typical of this kind of collaboration are projects in which engineers develop new technological approaches and social scientists (often social psychologists) are brought in at a late stage of development to ensure fit with social factors or to organise 'public acceptance'. The example of the collaboration between biomedical scientists and data scientists observed by Mao et al. also had features of this kind of collaboration: the biomedical scientists changed the research question partway through the project in a way that surprised the data scientists, and the latter had to take this new question on board and develop ways to address it. The data scientists were in service to the biomedical scientists, each group pursuing their work according to their disciplinary assumptions, but with the goals of some members taking precedence over those of others.

Sometimes this relationship is clear from the outset and made explicit in the roles, where one part of a team is labelled as 'research' or 'scientific staff' and the other as 'support' or 'professional' staff. A long-standing hierarchy in the sciences – that places natural science and engineers at the top – is an important determinant of whose agenda takes priority. Some disciplines are more powerful within the university or within corporations, often wielding more cultural capital and more resources (think engineers versus social psychologists or the sales department versus marketing). But sometimes the differences are subtler, at least at the beginning of a project. In such situations, there is often an imbalance in the perception of collaboration, where some members consider that they collaborate with others, but this perception is not reciprocated (A reports that they collaborate with B, but B does not report collaboration with A) (Zhang et al., 2020). Underlying this skewed perception is the way different contributions are valued as being substantive to the core objectives of the project or whether they are seen as non-essential, 'nice-to-have' elements.

The second paradigm in collaboration seeks to make the whole greater than the sum of the parts, and to question and go beyond the limits of established disciplines. In this paradigm, the disciplinary boundaries are not maintained. Models of collaboration in this paradigm seek to synthesise perspectives – not just results. Such collaborations have been heralded as promising a new kind of knowledge production (Weingart and Padberg, 2014) that may also be more open to lay knowledge in problem solving. In practice, this kind of collaboration involves the creation of common understanding of both processes and contents. This is sometimes described as the sharing of a 'trading zone' or 'third space', 'where people can compare, negotiate, and integrate goals, perspectives and vocabularies, as well as discuss shared meanings and protocols' (Mao et al., 2019, p.5) Ideally, this leads to shared criteria of quality and success. This is a challenging route (Mauthner and Doucet, 2008) that may feel risky to those involved. Such a mode of collaboration takes time and can feel uncomfortable or even unproductive. Whether it is possible to go for this kind of collaboration depends on the institutional setting, the kind of funding available and whether one is in a corporate or academic regime (or a mix). Yet, insofar as teams and team members can shape how they collaborate, it is possible to foster conditions that lead to better collaboration. Such learning and discomfort should be embraced as part of the job of pursuing fruitful collaboration. They can even be seen as the key to success. Teams should pay attention to creating the conditions for learning and to nurture learning in participants (Freeth and Caniglia, 2020).

Freeth and Caniglia offer the following suggestions. First, the creation of a common ground is important: it can mean sharing a concept or a method that is relevant to all involved. Creating a sense of safety and trust is also necessary. This can be done by making it acceptable to discuss failures in a team and to explore the diversity that may exist within a team. Spaces for interaction that do not reinforce hierarchies are also helpful (not feeling like some team members are 'guests' on the home territory of the rest of the team). Finally, teams should ensure that there is time to discuss both procedural issues (how to work together) as well as outcomes of the project (Freeth and Caniglia, 2020). Other factors such as personal affinity between team members and track record of participants also matter in developing effective collaboration.

Finally, from a more technical angle, there is a growing set of tools to support collaboration. Some tools are specifically aimed at data science collaboration, to support the documentation of the provenance of data and of code. Such tools, if well implemented and supported by a local culture, can contribute to keeping data work transparent and accountable. Other tools are directed to documenting data processing and analysis, such as GitHub, Slack or Jupyter Notebook. Of course, many aspects of collaboration are supported by more generic tools like email, document sharing and co-writing tools like Google Docs, and by file sharing services, as well as meetings, in person and online.

6.5 Conclusion: Becoming a data scientist today

While being a data scientist has been labelled 'the sexiest job' for almost a decade, what is associated with this role has changed. Data scientists were meant to help organisations capitalise on Big Data; later, to help personalise products or unleash 'intelligence'. Recently, data scientists have been seen as the key to achieving the goals of implementing machine learning and AI. Across these different kinds of hype, it has generally been recognised that being a data scientist means dealing with expectations from other parts of the organisation. Dealing with pressures and politics of organisations is yet another set of skills, and while they may be important for every job, when high hopes are pinned on establishing a new unit or project team, these expectations can be especially determinant. Besides the growing demand for data scientists, another significant development has been the 'mainstreaming' of data science. Many jobs in sectors not primarily associated with data science, such as education or policing, are increasingly requiring some awareness of data work and data science skills. There is also a growing value attached to skills and activities relating to the stewardship of data – this is not as yet implemented in most universities, but is strongly championed by funders and policy makers. For example, the job of 'data manager' is not yet well defined and those doing this job tend to be hidden under other administrative labels that do not do justice to their centrality in research. Overall, there seems to be increasing awareness that coding is neither the sole nor primary skill involved in doing data science.

People seeking to enter the data science job market will also find that it is strongly shaped by portfolio development and by participation in hackathons and other data competitions (Kaggle is probably the best-known brand). This means investing in demonstrating that you have particular abilities by sharing code or taking part in events – and providing free labour. This is quite different from relying on formal credentials, such as a degree from a recognised institution. It also means that networks and networking skills are especially important. This approach to recruitment tends to favour homogeny – people know people like themselves. So if job opportunities depend on who you know, there is little opportunity to diversify the workforce. This means that there should be increased attention to diversity of backgrounds of data workers, and particularly data scientists and data curators. Having only white, middle-class, technically trained men working as data scientists – as is still overwhelmingly the case in corporate data analytics firms – is far from ideal in terms of bringing a variety of experiences and viewpoints to the table. Going back to the questions we posed about how to handle binary sex classifications, it is clear that having data scientists with some knowledge of gender studies and an understanding of intersectionality would contribute to ensuring that the categories used for data classification do not discriminate or otherwise adversely affect relevant individuals and groups. Making positive efforts to address the current lack of diversity in terms of gender, age, ethnicity and class is a constructive step towards a better data science (we return to this topic in the concluding chapter of the book).

Finally, there are new civic and corporate roles for data scientists. The security and sensitivity of data, consequences and privacy concerns of data analysis, and the professionalism of transparency and reproducibility are all increasingly important areas of expertise in contexts beyond data science units in companies or research groups in universities. For example, data scientists have the expertise needed to work with or as journalists, to help make governments and businesses accountable because they understand how data is being generated and used. A recent investigation by the British independent daily newspaper *The Guardian* established that over a quarter of British councils (local government authorities in the UK) have invested in software contracts with large firms to support the administration of benefits to citizens (Marsh, 2019). The software systems are used in the activities of local councils. These include providing housing benefits (a subsidy for rental expenses), detecting signs of child abuse and allocating places to pupils in schools. Across these different applications, *The Guardian* reported concerns about privacy and data security, the ability of council officials to understand how some of the systems work and the difficulty for citizens to challenge automated decisions. In particular, the performance of some systems on 'predictive analytics' had been problematic: in detecting cases of potential fraud, low-risk claims for benefits had been wrongly labelled as high risk. This case of too many false positives had harsh consequences, delaying the payment of benefits and causing undue hardship for vulnerable groups (Marsh, 2019). This is an example of how social and political issues require an understanding of data science in order to be tacked. Many NGOs and citizen movements are therefore drawing on the expertise of data scientists to navigate and evaluate the use of large-scale data systems, whether to create alternative datasets (citizen sensing projects) or to audit and demand more responsible systems.

As machine learning and data-driven policy spread across different levels of government and areas of life, the need for such experts will also increase – data scientists can be heroes of social justice, as well as the champions of new business models and drivers of novel scientific insights.

ADDITIONAL READING

Edwards, P. N., Mayernik, M. S., Batcheller, A. L., Bowker, G. C. and Borgman, C. L. (2011). Science friction: Data, metadata, and collaboration. *Social Studies of Science*, 41(5): 667–690.

Jemielniak, D. and Przegalinska, A. (2020). *Collaborative Society*. Cambridge, MA: MIT Press.

Meng, X.-L. (2019). Data science: An artificial ecosystem. *Harvard Data Science Review*, 1(1). https://doi.org/10.1162/99608f92.ba20f892

Wouters, P., Beaulieu, A., Scharnhorst, A. and Wyatt, S. (eds) (2013). *Virtual Knowledge: Experimenting in the Humanities and the Social Sciences*. Cambridge, MA: MIT Press.

7

GOVERNANCE OF DATA JOURNEYS

———————————— Overview of chapter ————————————

Summary

The final chapter of this part examines the governance of data journeys. We consider a number of questions: Why do we expect that some kinds of data can travel and others not? Where can these data travel, and how is travel shaped by regulations, laws and socio-technical systems? How can diverse contexts of use affect the journeys of data? To address these questions, we first introduce the notion of data governance and discuss two broad modes of governance: Closed Data (including proprietary rights and security and privacy regulations) and Open Data. We identify and illustrate the difficulties linked to these modes of governance by examining the difficult case of biomedical data. We then consider the implementation of **FAIR** principles, which carves a 'third way' for data governance focused on Usable Data, and thus offers a potential solution to the difficulties posed by Open and Closed Data. Finally, we discuss how data travel across national borders and we highlight the problem of data inequity. This problem is a basic social concern that FAIR principles do not help to address. We close by stressing that while data governance is a crucial component of data work, it is not possible to fully capture it through rules and regulations, no matter how elaborate. To choose and implement an adequate form of data governance requires in-depth engagement with the specific characteristics and conditions of data journeys.

7.1 Introduction: What is data governance?

What role does access to data play in data science and related types of work? Under which conditions is it appropriate and useful to share data, and with whom? What measures can and should be used to oversee and, where appropriate, support data travel across the globe? What constitutes appropriate reuse of the data being released, and how can this be monitored and incentivised? These questions are sometimes set aside by those engaged in everyday data work. It is often assumed that there must be legal and regulatory norms in place to oversee and constrain the handling of data and its implications. As long as such norms are respected, then there would be no need for people on the ground to actively engage with such issues. All that is needed, it is sometimes argued, is to follow existing rules or ask relevant authorities for further guidance when rules are lacking. In this chapter, we demonstrate that more engagement with these issues is needed for data work, both because data governance is more than a system of rules, and because it is often not clear who the 'relevant authorities' are who can provide guidance on how to share and reuse data. In fact, everybody who is involved in the collection, curation and analysis of data needs to actively engage with questions around who can access and use the data, how and why.

Making this argument involves taking a close look at **data governance**, which is the ensemble of regulations, norms and socio-technical systems that enables and directs data work – and particularly how, where and why data can travel. We show that data governance plays a fundamental role in shaping what can and cannot be done with data, and whether they can be accessed and reused in the first place. One obvious way in which this happens is through legal regulation, such as, for instance, the laws around data protection. These laws constrain what forms of data work are admissible, in which context and for which purposes. Acquiring an understanding of existing regulations around data privacy, security and protection, as well as proprietary rights and regimes attached to data, is therefore a crucial step towards any form of data work. Adherence to rules and compliance with existing legal systems do not, however, replace active, critical engagement with the conditions for data travel and reuse. In other words, following the law is not enough to ensure effective data governance in any one situation. There are three main reasons for this.

First, data governance includes not only the law, but also the informal principles and rules that are socially agreed to apply to data work, such as for instance whether or not data should be owned and traded, and what constitutes reliable and valuable data. Data governance also includes the socio-technical systems used to carry out data work, such as the specific platforms, infrastructures and forms of agency through which data are produced, stored, disseminated and analysed. In what follows, we will examine three broad forms of data governance: Closed Data, Open Data and Usable Data. As we will see, this is an oversimplification of reality, since each of these forms of data governance encompasses a multitude of social systems. Indeed, data governance systems are necessarily flexible and ever-changing, since they need to adapt to specific socio-political and economic contexts, they serve different goals for different people and they are intertwined with other systems in sophisticated and diverse ways. Nevertheless, distinguishing between these three forms of data governance will help to highlight their advantages and disadvantages, as well as the fact that none of them provides a perfect solution to the multiple social, economic, cultural and political concerns raised when making data travel. Hence our distinction between Closed, Open and Usable Data is not intended as a comprehensive summary of all forms of data governance in the world, but rather as a starting point for you to learn to recognise and disentangle aspects and implications of data governance, and their relevance to your own interest in data work.

Second, because data governance can take different forms depending on its socio-political setting (including the institutions and governments that enact it), data workers often need to deal with different interpretations of data governance at the same time. For instance, researchers working for an international private–public partnership are subject to their university's interpretation of data governance as well as to the interpretation used by their industry partners, which may well be quite different. Researchers working in multinational projects may similarly be subject to very different understandings of data governance by the various governments that are sponsoring their work. This creates a real conundrum for data workers: Which interpretation of data governance should they follow, when and why? And to whom are they ultimately accountable? Answering these questions requires critical thinking and the ability to decide and justify a specific course of action.

Third, even in the rare cases where the regulations used as part of data governance are clear and consistent, they do not dictate every aspect of data work. Technologies, methods and opportunities surrounding data use evolve too quickly for any rule-based or legal structure to keep up, and the most detailed set of rules cannot control all decisions taken in the course of a data-intensive project. There comes a point where data workers need to decide for themselves whether a given decision complies with existing rules and conforms to the expectations underpinning the system at hand. Moreover, researchers working with personal data need to understand that for the decisions they take, they can be personally required to demonstrate compliance with data protection regulation and the rights of **data subjects**. When it comes to everyday technical decisions around how to collect, format and share data, data workers therefore need to be able to think strategically about the broader implications of their methodological choices, and be accountable for how and why their choices enact specific forms of data governance.

7.2 Data as private commodities: Closed data

In Chapter 2 we discussed how some scientific fields have a long tradition of sharing data, for example in the areas of climate science or genetic research. This history demonstrates that moving data around is not so simple. This is because data are not only significant as research components, but often also as the result of considerable financial investments and/or personal efforts, as opportunities for communication, as currencies for exchange and as tokens of personal identity. Think about the decades of work and substantive funding required to build the gigantic particle accelerator hosted by the European Organization for Nuclear Research (CERN) in Geneva, which made it possible to produce data used by particle physicists around the world. Another example could be the millions of dollars spent every year by the pharmaceutical industry to set up clinical trials, thus obtaining data documenting the effectiveness of specific medical treatments. These are clear instances of data as investment. Data produced by users of social media are good examples of the role of data as tools for communication, which can in turn also function as currencies of exchange. When Facebook sells data produced by its users to companies interested in using such data to understand consumer behaviour and preferences, or when data produced by a person's interactions with the internet are used to profile that person and define their 'digital self', then data have respectively monetary or personal value.

These examples show that data can be endowed with different forms of value depending on their context of use, an idea that we return to in more detail in the next chapter. In this section, we focus specifically on the governance implications of the value of data as the result and object of investment – in other words, the fact that data are widely regarded as sought-after commodities, and how this fact shapes the travels of data and the rules and regulations set up around their reuse. While data have long been part of negotiation of scientific, political, social and economic interests, they have been increasingly seen as intrinsically valuable assets in the current millennium. Acknowledging that all types of data hold commercial potential, especially when aggregated to analyse and predict human behaviour, is key to understanding the role played by data in contemporary society. A glance at the most rapidly growing and successful businesses, both nationally and internationally, over the past decade suffices to reveal that businesses and start-ups focusing on data analytics and data security have grown exponentially, along with legal services focused on intellectual property rights over related technologies and algorithms and data licensing. Nearly all sectors rely on assistance from firms and consultants specialised in data work – from planning political campaigns to launching new products or businesses, auditing the effectiveness of any service and monitoring health. The collection, mobilisation and analysis of data is certainly not restricted to the world of scientific research or public government. Data work has become a basic function of the capitalist-imprinted economic development championed by the global free market.

The view that data as a private commodity can be bought and sold is very prominent. This prominence has implications for the governance and mobility of data. Nowotny et al. (2001) have observed that data is actually moving less freely than in earlier periods, partly because knowledge is no longer seen as a public good, but rather is seen as intellectual property 'which is produced, accumulated, and traded like other goods and services in the knowledge society'. Not only is data treated on a par with any other commercial good and subjected to the dynamics of free market, but the commerce of data and the growing awareness of its key role in the capitalist economy have significantly stimulated its transit via digital platforms (Thrift, 2005; Beer 2016; Srnicek, 2016). Companies such as Google, Apple, Facebook and Amazon have grown at breakneck speed to become some of the richest and most powerful corporations in the world. A growing proportion of the resources required to collect, store and analyse Big Data is therefore under the control of institutions with mainly commercial interests both in the public (government) and in the private (corporations that are active in research) sector. This means that there are fewer opportunities for those who have little financial or social power to take part in the building of tools and strategies for analysis and interpretation. In other words, we are seeing the establishment of an information and knowledge production regime centred on the privatisation and commodification of data.

The success of data as globally recognised commodities is accompanied by sophisticated ways to control the flow of data and to regulate their use. There are also many measures taken to ensure that the interests of those who invest in data work are protected. For example, data produced in the course of a commercial operation, such as a social media platform, are stored and organised so that access to them is restricted. Data infrastructures are built to be highly secure and to make strict monitoring of data access possible. This use of data infrastructure is an example of how socio-technical systems implement data governance. In this case, the infrastructure is built to control **data mobility** and thereby implements what we here call Closed Data governance. In addition to infrastructure, contractual agreements contribute to data governance. The movement of data is the object of agreement between data subjects (in this case, the users of social media) and the company that collects the data (the social media platform), where the company states the conditions under which data will be accessed and reused, and subjects agree to such conditions. In turn, these agreements need to comply with governmental regulations and data protection laws governing commercial activities within the country in which the company is based. Moreover, companies can decide to monetise the data by selling them to prospective buyers, which again requires legal agreement between the parties involved about who retains ownership of the data and what ownership means in terms of data reuse. Companies can and must also decide whether the transfer of data from one company to the other complies with data protection laws and the original contract that the social media platform made with the users of the platform. It is these kinds of technical, social, legal and contractual arrangements that shape decisions about which data are made available, to whom and under which conditions, and which data are kept secret or confined. The approach chosen in many data protection laws is to give rights to individuals whose data are being collected and reused (often referred to as 'data subjects'), like the right to be forgotten and the right of algorithmic transparency. Individuals who reside in countries subject to those laws can exercise these rights, even if they gave consent to use the services. The case of Maximilian Schrems against Facebook is considered a landmark case to demonstrate that these rights can affect the governance of Closed Data (Kuner, 2017).

Within this system of data governance, the circulation and sharing of data are thus subject to contractual agreement between individual parties, as well as the legal and regulatory restrictions that apply to all forms of commerce. This means that data are conceived as private property that cannot be freely shared and only travel as a result of financial or other compensation. Widespread sharing of data of relevance to business activities is often considered to be risky because it might lead to losing competitive advantages. Yet, there can also be a downside to having strict restrictions. Consider the evolution of the governance of Twitter data over the past 10 years. When Twitter was born, the company had a markedly permissive attitude towards access and reuse of its data for publicly funded research or innovation led by start-ups. This changed as the market for social data has grown. Since 2011, Twitter application programming interfaces (APIs) have gradually restricted access to free-range data uses. This includes activities of the third-party developer community that is arguably responsible for many of the innovations that give Twitter its unique culture. Third parties are responsible for innovations like the @reply, the hashtag and trending topics. Under the current governance of Twitter, it is difficult to imagine how these innovations could have taken place. The current governance also restricts research, since privileged access is now only available to paying customers (Bruns and Burgess, 2016). Such restrictions that favour commercial use of data as private commodities are an example of Closed Data governance.

Closed Data governance can also be found in the handling of data that are considered sensitive for other reasons than their commercial value. This is the case when data of military or political relevance are linked to security concerns. Government data about military operations and facilities, and military-related contracts, for example, are often kept hidden from public scrutiny through a combination of rigorous data protection laws, rules preventing data workers from disclosing anything of relevance to their jobs to outside parties, secure digital infrastructure and active policing of hacking attempts. Reasons for this secrecy include preserving national security, diplomatic confidentiality and the integrity of political processes and expert advice about sensitive situations, and avoiding undue interference and misunderstandings over governmental decisions. Breaches of Closed Data governance in such cases can have momentous implications, as demonstrated by the political fallout of Edward Snowden's decision to disclose classified data held by the US National Security Agency in 2013. This disclosure prompted a global debate on the methods used by surveillance agencies to obtain intelligence on their targets. Another good example is WikiLeaks' publication of confidential documents of American public officials in 2015 and 2016. Among many other confidential materials, the leak included personal email correspondence by Hilary Clinton, whose contents played a significant role in the failure of her presidential bid in 2016. Closed Data Governance is therefore made up of different elements that include regulations (laws, rules), contractual agreements and socio-technical infrastructures. This mode of governance favours certain uses of data, ensures who will benefit from data, and restricts some forms of research and accountability.

7.3 Data as public goods: Open data

Despite its prominence in contemporary society, the idea of data as private commodities is not uncontroversial. For instance, many data studies scholars have argued against the treatment of personal data like a home address or date of birth as commodities. They claim that such data constitute public information that cannot be owned by anyone, not even the subjects on whom it was generated. According to this line of reasoning, since it cannot be owned, it should not be bought and sold either. Similar arguments have been put forward for environmental, agricultural and climate data, insofar as they are relevant to the whole of humanity. Such data, the argument goes, can and arguably should be used for the benefit of all. These arguments stress the moral economy of data (Strasser, 2019), and insist that the potential of data to improve life on the planet needs to prevail, and that there is an associated responsibility to ensure that such value is realised.

A popular understanding of data that embraces these ideas – and stands in opposition to a market-driven view of data as private commodities – is the understanding of data as public goods. This is typically associated with a view of data as 'non-excludable' resources. This means that the use of data cannot be the privilege of specific individuals or groups. It also highlights that data are 'non-rivalrous' resources, in the sense that they can be reused as much as needed without becoming depleted (Hess and Ostrom, 2007; Borgman, 2015). In this view, data are expected to be freely and widely available to anybody who wishes to use them. This represents what we shall call an open mode of data governance: the development of infrastructures, venues, laws and regulations that support data journeys while imposing the least possible constraints on who uses the data, why and how.

One place where Open Data has made considerable headway as a system of data governance, in contrast to the system of Closed Data described above, is the world of research data (Leonelli, 2013; Borgman, 2015). Within domains such as meteorology and astronomy, there is a long history of international data sharing that demonstrates the opportunities that arise when pooling data work across disciplinary and national borders. These include the opportunity to accelerate the pace of discovery through collaboration; to avoid useless duplication of efforts and investment; and to increase the transparency and accountability of research by widening the peer group available to scrutinise results, address mistakes and prevent fraud. Other opportunities include improved dialogue and exchanges between professional researchers and other groups who may have crucial expertise to offer on the topics being investigated (i.e. when engaging patient groups and physicians in biomedical research on specific diseases).

Indeed, in some areas of scientific research data are conceptualised as a **knowledge commons**. This is a public good that contributes key insights on human life and therefore needs to be accessible without restrictions. A well-known instance are genetic sequence data, which were released free of charge in 2000, following a debate around whether or not such data should be privatised, given their potentially immense commercial and humanitarian value. Such instances of Open Data governance have acquired increasing support as a central component of **Open Science**, a movement committed to promoting collaborative research practices and the widespread distribution and reuse of data, results and methods. Open Science aspires to take advantage of new communication technologies to enhance access to and use of all the components of research, ranging from data to models, software, techniques, instruments, protocols and review procedures (Boulton et al., 2012; Fecher and Friesike, 2014). In 2016, the European Commission endorsed Open Science as crucial means to foster research excellence and impact by increasing scientific transparency, accountability and reproducibility (European Commission, 2016). As part of this commitment, it also invested in the creation of a European Open Science Cloud, a federated system of databases aiming to provide a common platform to store, manage and access research data produced across the European Union. By 2019, many countries including France and The Netherlands launched National Open Science Plans, including support for national infrastructures and cloud systems, while the United Nations (UN) flagged Open Science as the 'core enabler' of their 2030 agenda (United Nations, 2018; Beaulieu, 2021). Open Data was also backed by many funding agencies, including both public entities such as National Science Foundations and Research Councils and private funders such as the Wellcome Trust and the Bill and Melinda Gates Foundation (Burgelman et al., 2019). The involvement of these institutions meant that researchers following this model are now incentivised and rewarded in the form of access to funding, and there are systems of oversight over how funded projects manage data to ensure Open Data governance. These shifts and associated infrastructures make it possible to consider alternatives to rival the closed circuit of data privatisation.

The rise of Open Data governance is far from affecting only the world of research (Kitchin, 2014). Its reach extends to government and social services (through the idea of 'Open Government') as well as private industry and particularly large corporations. In the area of biomedicine, the emerging movement for 'Open Pharma' is especially important given that pharmaceutical research has been one of the biggest proponents of Closed Data governance in the past decades. The arguments in favour of Open Data governance stress the fruitfulness of Open Data for knowledge development, social inclusion and public accountability. Across these sectors, the idea that free data circulation is a highly desirable, profitable and effective way to handle data has gained considerable traction. Many national research agencies and international organisations such as the European Commission, for instance, expect that openness will involve more participation, a more fluid and effective division of labour between data producers, curators and users, better feedback and scrutiny leading to more reliable data resources, and more reuse of data because it is easier to source them. In turn, this could hold advantages for companies, since opening up the data could help companies to outsource data work and provide a commercial advantage over costly in-house data work.

7.4 A hard case: The journeys of health-related data

The history and contemporary practice of data work is underpinned by the unresolved tension between conceptions of data as public goods and private commodities. Both views cast a long shadow over the development of data practices and technologies and exert a powerful influence on debates around how to mobilise specific types of data. In practice, this often results in complex governance arrangements that are neither fully open or nor fully closed.

As an example, consider the debate over the ownership and dissemination of genetic sequence data that took place at the turn of the millennium (Jones et al., 2018). This prolonged and acrimonious controversy culminated in a clash between the then-director of the National Institute of Health, representing the consortium of public funders that sponsored the Human Genome Project, and the head of Celera Genomics, a private company that invented a technique to speed up the sequencing process. Celera Genomics hoped to retain an exclusive right – at least in the short term – to use the data, so as to offset its investment in the technology and to reward its creative approach to a difficult technical problem. This attempt to privatise genetic sequencing data did not succeed however, and arguments for the significance of these data as public goods prevailed. Concretely, this led to the swift publication of the data and a subsequent agreement (known as the 'Bermuda rules') requiring geneticists to donate data – and particularly non-human genetic sequences, which are not subject to privacy concerns – to public databases. This dispute set an important precedent for Open Data practices and their potential impact. It also made possible countless subsequent discoveries that were at least partly due to the ease with which researchers could access and share existing sequencing data. Yet, the dispute also underscored and exemplified the financial advantages of owning genomic data, the significance of finding ways to monetise and reward biomedical research, and the ethical and legal problems involved in sharing human genetic data. These lessons were learnt by commercial sequencing companies such as PatientsLikeMe and 23andMe. These private companies developed a service that offered to sequence their clients' genomes and interpret results securely and cheaply – in exchange for the right to retain and use the data, in an anonymised form, for their own research purposes, thus effectively closing down access to the data (Tutton, 2016).

There are many other examples where arguments for Open Data intersect with proprietary rights, as well as security and privacy regulations. The boundaries between Open and Closed Data governance become blurred. Some scholars have even argued that Open Data governance is a contradiction in terms, since it depends for its very existence on understanding of data as a piece of intellectual property that needs to be managed accordingly (Wessels et al., 2017; Mirowski, 2018). In other words, it may not be possible to conceive of data as public goods without conceiving of them as commodities, and vice-versa. For instance, it is not uncommon for Big Tech companies such as Facebook and Google as well as the hundreds of companies set up in the past decade to enable the acquisition and sale of data for commercial objectives, to defend two apparently opposing ideas at the same time. On the one hand, they argue for the idea that personal data are largely public information as they are easily accessible (such as name, surname and address), and that they are therefore reusable for any purpose once people have provided their consent. On the other hand, they also put forth the idea that personal data, if not necessarily private, can at least be 'privatised' and therefore bought and sold like any other product. In such a case, the confusion that reigns over the meaning of data 'ownership' is used to encourage individuals to give up rights over their data to a growing number of companies, often on the basis of gaining access to services that make their daily life simpler (such as information on traffic, the nearest cinema and the weather forecast for the weekend).

In this section, we explore such problems by considering the specific case of health-related personal data, whose journeys raise serious concerns with both the Open Data and the Closed Data forms of data governance we have considered thus far. This is because of the sensitivity associated with personal health data and because of the field's history and ongoing intellectual trends. These include debates over what methods of data generation and interpretation are viewed as reliable and which types of expertise should be involved in clinical assessments and interventions. In the biomedical sector, there are emerging trends to assemble and link diverse types of data through digital technologies. Among those, most notable are *personalised and precision medicine*. They target research and therapeutic intervention to specific individuals and groups. This goal can only be achieved by consulting a vast and diverse body of evidence. Such evidence includes data collected through lab studies of non-human models, data provided by patients through social media, data generated through direct-to-consumer genetic services, data obtained through clinical and longitudinal studies on various human populations, and data produced within clinical sites such as hospitals and individual GP practices (Hood and Friend, 2011; Vayena and Prainsack, 2013; Kallinikos and Tempini, 2014; Merelli et al., 2014; Lucivero and Prainsack, 2015; Green and Vogt, 2016). Added to these diverse sources are data produced through self-tracking; that is, the use of measurement technologies enabling individuals to collect data about their own physiology, behaviour and activities, including information such as blood pressure, heart rate, and intensity and regularity of physical exercise (Beer, 2016; Lupton and Michael, 2017). The collection of such data is sometimes associated with the dissemination of personal data through social media, either as a form of social engagement with family and friends (the 'sharing activity' option offered by Apple Health) or as a contribution to biomedical research (as in the above-mentioned case of PatientsLikeMe) (Tempini, 2015; Sharon and Zandbergen, 2017). Depending on the legal framework and user agreements in place in each case, health-related data collected through self-tracking tools may be acquired by data analytics companies and become part of privately owned Big Data collections (Ebeling, 2016). Similarly, many companies that provide direct-to-consumer testing services (e.g. in genetics) tend to keep results within their in-house databases (Harris et al., 2016).

In all such cases, to make it at all feasible for data to travel across contexts and thus inform health-related research, all stakeholders involved (whether they are patients, market structures, research institutions or policy bodies) need to acknowledge and negotiate the value of data as political, financial and social objects. This often generates tensions around who owns the data and what constitutes acceptable reuse, with legal and ethical concerns at the centre of ongoing debate around data work in this area. That the sale of data – especially of personal data – can have negative consequences for individuals and communities is becoming clearer, even to those who are not directly involved in this type of work. In Europe, an extensive legal framework called General Data Protection Regulation (or **GDPR**) was introduced in 2018 to protect individuals from abuse of their personal data and encourage the development of sophisticated and responsible ways of archiving and mobilising data. We highlight a few important features of GDPR and recommend consulting the webpages of the EU for more detail on GDPR and its workings. GDPR builds on extensive documentation by international organisations such as the Organisation for Economic Co-operation and Development (OECD) and the UN, as well as data protection legislation at the national level, which has long existed in many countries. Despite such laws and regulations, however, there is at present no internationally recognised framework specifically targeted at health data. Furthermore, the application of such laws in the biomedical domain is fraught with ambiguity (OECD, 2017). This is also because data protections laws allow exceptions when data are used for research geared to the public good, such as biomedical research. In those cases, safeguards and appropriate measures to protect the rights of data subjects may be determined by discipline-specific ethical codes and relevant regulation, as well as researchers' own determination of what constitutes acceptable data use.

More broadly, there are several conflicting ways of interpreting what may constitute 'public interest' and 'common good' in the case of Big Data sharing and reuse for medical purposes (Floridi, 2014; Nuffield Council on Bioethics, 2015; Brunton and Nissenbaum, 2015; Prainsack and Buyx, 2017). It may well be that data disseminated to foster biomedical research end up causing harm to individuals or groups, or that such data are misappropriated and misused for purposes other than research. In such cases, governments and research institutions need to take responsibility for protecting individuals (and particularly patients) from harm, whether or not those individuals have decided to exercise their right to participate in scientific research. This is particularly important in light of the commercialisation of personal data described above, and the vast potential for fraud and manipulation. Given the ease with which individual data points can be aggregated and anonymisation procedures can be reversed, confidentiality and patient protection remain very important in medical research practice (Hogle, 2016). It is also argued that privacy frameworks need to be extended to groups and local communities. Imagine a case where local hospital services are tailored to the symptoms and illnesses that Twitter users living in that area complain about on the platform (e.g. Floridi, 2014). This analysis of a combination of personal and geolocation data from platforms could affect whole communities. As recognised by the OECD in its latest report on the governance of health data (OECD, 2017), patients' privacy needs legal and social protection like never before.

Within the public sector, worries around security and privacy have shaped data management, and in some countries, they have created a risk-averse data culture. The result is that data are strongly contained and access restricted (in 'data silos') and rarely shared across departments and agencies. In the UK, for example, the National Health Service has long failed to implement socially responsible ways of integrating data about patients, resulting in a public backlash against data sharing more generally. This has made it hard for medical specialists and physicians to exchange information on a single patient. As a result of this lack of flow of information, patients face fragmentation of care. This is especially problematic for patients with complex conditions who need consultations with several specialists and oversight by their general practitioner (as became particularly evident at the outset of the coronavirus pandemic). The high degree of fragmentation of medical data services in the UK also translated into highly fragmented data storage, with several unconnected IT systems involved in maintaining data collections from patient groups, clinical studies and other biomedical research.

The case of biomedical data is a striking example of how difficult it is to devise and consistently apply a system of data governance that results in effective and fruitful data work. Similar issues apply to other domains too. Geolocated data, for instance, are often reliant on satellite data, which may be public or private, and on proxies such as triangulation of mobile phone signal or IP addresses of internet-connected devices. The network operators that collect and control the latter data are overwhelmingly private companies. The language of Article 76 of the UN document on the Sustainable Agenda indicates great optimism that this can be addressed through 'appropriate public–private cooperation to exploit the contribution to be made by a wide range of data, including

Earth observation and geospatial information, while ensuring ownership in supporting and tracking progress' (UN News, 2016). The effectiveness of such cooperations remains doubtful however (Van Dijck et al., 2018; Zuboff, 2019).

7.5 Shifting focus to usable data: The FAIR principles

The previous section provided a sense of the complexities associated with data governance, and the difficulties of putting data to work when confronted with very different socio-technical systems with diverging priorities and understandings of how data should travel. How such clashes are negotiated and resolved shapes the travel of biomedical Big Data and the effectiveness with which they are used to produce new knowledge. Perhaps most importantly, negotiation around these clashes shapes what kind of knowledge is extracted from the data in the first place. What has made data so crucial in contemporary biomedicine is precisely their multifaceted perception as at once local and global, free commodities and strategic investments, common goods and grounds for competition, potential evidence and meaningless information. If data are to travel across academic labs, industrial development departments, policy discussions and social media, those data need to be of interest to all actors involved. But, as we have seen, motivations and incentives behind efforts to collect, share and reuse data diverge widely. Is it possible to retain some flexibility to multiple uses and future scenarios, as well as include the divergent interests of potential users in a form of data governance? Such a data governance would help data workers deal with possible conflicts among data values in their work, thus balancing the constraints and decisions internal to their operations, and the broader landscape of opportunities, demands and limitations within which they operate.

Over the past few years, another form of data governance has emerged precisely to enable a flexible approach to diverging interests, while at the same time ensuring that data are used in ways that are broadly acceptable to all relevant stakeholders. A group of prominent data experts, many working within biomedicine and other academic research domains, argued that data put online do not need to be unconditionally accessible to all, as in the case of Open Data. Rather, data need to be:

- Findable, in the sense of being easily searchable and retrievable;
- Accessible, in the sense of being available for consultation;
- Interoperable, in the sense of being formatted and curated in ways that make it possible to link with data from other sources;
- Reusable, that is having enough accompanying information (e.g. metadata) that enables their validation and analysis.

These four 'FAIR principles' (Wilkinson et al., 2016) have been well received and widely adopted by scientific agencies, governments and some companies, including the European Open Science Cloud. The FAIR principles constitute a pragmatic approach to enable data reuse, even when the data themselves are not open. All that is required is that data can be located and the procedure through which they can be accessed is clearly and explicitly laid out. This means that whether or not a prospective user has the right to access data depends on their purposes and the implications of data release for data subjects (a view that aligns well with the relational view of data discussed in Chapter 4). In this sense the FAIR principles are recognised as useful guidance for data management and sharing. They have systemic implications for the processes, technologies and institutions used to make data travel. We refer to the socio-technical systems associated to the implementation of FAIR data as Usable Data governance. This is a form of data governance that attempts to carve a 'third way' between open and closed regimes of data governance, by making data access and reuse subject to specific conditions set up by data providers. Prospective data users thus need to understand and agree to these conditions in order to be able to work with the data.

Making data FAIR clearly requires making data mobile and useful as evidence across sites, contexts and uses. As we saw in previous chapters, this requires creating and maintaining trustworthy, user-oriented data infrastructures. It also demands the development of specific skills and expertise in data management and curation. In response to these demands, fields such as data science, bioinformatics and biocuration are rapidly acquiring prominence alongside traditional scientific disciplines. Large corporations such as Google and IBM are positioning themselves as providers of data analytics and data enrichment tools. Within academia, the expertise of those who produce, curate and analyse data is increasingly acknowledged as indispensable to the effective use of Big Data. This encourages significant shifts in governance and traditional hierarchies; the importance of laboratory technicians, librarians, administrators and database managers in knowledge production is becoming clearer and more visible.

The same goes for ideas about research excellence. Some national funders move away from research evaluations based solely on measurements of the impact of scientific publications. Research agencies in The Netherlands, Finland and Slovenia, for instance, are shifting from citation counts and journal impact factors to alternative measures that also emphasise the scholarly commitment to sustainable data management (Wilsdon et al., 2017). This in turn forces scholarly publishers to reassess their business models and dissemination procedures, and research institutions to adapt their administration to this new landscape. These shifts are supported by policy bodies and research funders like the European Commission, the US National Institutes of Health (NIH), the Wellcome Trust and the Bill and Melinda Gates Foundation. All these major funding institutions view Usable Data as a step towards enhancing scientific excellence and public trust in science, and they lead efforts to foster the dissemination and reuse of data generated through public and private research.

7.6 International data journeys and the problem of data inequities

While Usable Data governance improves the conditions for data work, and helps to break the stalemate between open and closed approaches, it fails to address a major social concern underpinning most data journeys and use scenarios. Usable Data governance does not make a difference in the extent to which data journeys and related technologies reinforce existing inequities and widen the gap between rich and poor.

In government as much as in academic research and industry, big players with large financial and technical resources lead the development and uptake of data analytics tools. This puts the rest of society at the receiving end of innovation in this area. This means that the priorities of big players determine which new activities, products and services are developed. All too often, users of digital technologies are conceptualised as customers with specific financial and social characteristics and abilities. Social life is increasingly filtered by digital technologies, including access to education, social services and political agency. As a result, individuals who do not possess adequate hardware (a smartphone), infrastructure (access to a stable broadband connection), skills (the ability to choose and manage apps) or status (a clear legal identity as a citizen of a specific country, including the right to set up a bank account) become socially marginalised. They lose (or never gain) the ability to make themselves seen and heard in the digital sphere; nor do they have a say in how this digital sphere is shaped. Such individuals are often already at the margins of society, especially when they deviate from expected norms (as in the case of differently able, elderly or chronically ill people or asylum seekers), when they are socially isolated and/or experience financial hardship. Thus, contrary to popular depictions of the data revolution as a harbinger of transparency, democracy and social equality, the *digital divide* between those who can access and use data technologies and those who cannot continues to widen. Access is only part of the problem, since, as mentioned above, *what* can be accessed or not is also an important issue. Having a say in what data, tools and services should be developed is also fundamental to creating fair data governance. Participation and representation in decisions about data use and services as well as in decisions about priorities in data work are fundamental.

Such exclusions are pervasive and visible in all countries, no matter how well resourced, and irrespective of economic conditions. They have profound effects on what forms of data use are actually performed and what data are produced in the first place. While the vast majority of the population is encouraged to provide more and more personal data for access to digital services, many encounter significant barriers to contributing data and have no control over what data is produced about them. The use of personal data by insurance and credit companies is a case in point. These companies regularly combine a variety of data sources on their clients in order to produce a customer profile that will be used to decide which services will be made available to that specific customer, and how much to charge them. However, and despite the sometimes severe economic and social implications of such decisions, customers of such companies rarely have access to the data used to produce such profiles, nor can they check the criteria and parameters against which they are being profiled.

Moreover, many do not have the means to consider the multiplicity of uses to which such data can be put and the potential for positive or negative repercussions for themselves and their communities. This results in potentially unfair modes of participation in data collection and analysis, with some social groups being represented more heavily than others depending on the domain, and little protection from their resulting visibility (or lack thereof) as research subjects and the claims derived from the analysis of such data. For instance, police forces tend to collect more data about individuals from visible minorities than they do about white people. Public consultations on social services tend to attract more responses from middle-class citizens than from low-income and/or disadvantaged citizens and from those who, precisely because they are not citizens, need social assistance the most.

Within the world of research, another type of discrimination concerns the extent to which data are accessible and reusable to researchers and other communities around the world. Differences in access also shape the kinds of approaches, disciplines and locations that conduct Big Data analysis. Research databases mostly display the outputs of rich, English-speaking labs within visible and highly reputed research traditions that deal with 'tractable' data formats (such as genomics data in biology), a phenomenon that has been analysed as a linguistic divide (Amano and Sutherland, 2013). The involvement of poor/unfashionable labs and researchers working in low-income countries is low and almost always at the receiving end (meaning that they are not involved in developing resources, just consulting them). The issue of Big Data access compounds existing digital divides locally and internationally with a new divide in access to data as well as appropriate technology, resources and trained personnel to be able to reuse such data. Furthermore, the resources required to collect, store and analyse Big Data are increasingly being appropriated and developed by a few multinational corporations and governments, with little opportunity left for less powerful and internationally recognised players to participate in shaping the relevant technologies and strategies.

In the private sector, it is also unclear what the status of data is, and which data are being shared with whom. This is especially the case in secretive industries with a strong commitment to Closed Data governance such as pharmaceutical companies and agro-tech corporations. This divide between those who have and those who don't have the capacity to become involved in Big Data usage has severe implications for researchers based in low-resource environments, with inequalities in visibility, power and location being reinforced, rather than mitigated, by Big Data dissemination (Bezuidenhout et al., 2017). The divide also results in the scarcity of data relating to certain subgroups and geographical locations, and thus to questions around whether individuals who remain excluded from Big Data collections benefit from advances such as personalised medicine and precision agriculture. This limits the comprehensiveness of available data resources, restricts the potential for Big Data use to tackle global challenges and constitutes an additional source of potential bias – an issue we return to in the Data Story on Sustainable Development Goals in the next part.

Thus, existing inequities in the world are being widened, not diminished, by existing data infrastructures and related regimes. Attention to Usable Data and governance structures can go some way towards addressing this issue, for instance by improving the accessibility of data and ensuring that data can be reused by those who may need them. However, all of the data governance systems analysed in the previous sections can still create huge inequities. Making data open does not ensure that they are accessible and usable – that depends on the training, technologies and settings available to potential data users. Keeping data closed is obviously problematic in relation to digital divides, particularly since control over Closed Data is often retained by powerful, wealthy actors such as international corporations or national governments. And appeals to FAIR principles, while tackling questions of data accessibility, do not directly address questions such as who should benefit from data reuse, how can accessibility be ensured across highly diverse user scenarios and how can data misuse be prevented. To address such issues, one needs at least three factors: a legal framework for data reuse, including laws against discrimination; lobby groups and international organisations devoted to improving conditions for data access, production and use for all in ways that are fair; and a firm commitment by data workers themselves to help tackle these concerns.

7.7 Conclusion: Governance is not a silver bullet

In this chapter, we have shown that data governance, in the form of regulations and socio-technical systems, shapes the journeys of data and the conditions under which they travel. Data governance is fundamental to the mobility and usability of data. At the same time, we argued that no single form of governance is perfect, nor can governance in and of itself be a silver bullet that solves all the difficulties and social concerns linked to the travel of data. As we discussed, even in the case of Usable Data governance, straddling open and closed systems, there are unforeseen negative impacts on society that cannot be fully addressed through legal and technological improvements.

One element of governance that we have briefly mentioned in this chapter, without however delving into details, pertains to the ethics of data work, and related legal and scientific regulation to ensure that data are treated and analysed responsibly. This is a crucial issue for data governance, on which we will focus closely in Chapters 9 and 10. The lessons learnt in this chapter, around the significance of data governance and also the difficulties of formalising systems of rules and regulation that can offer concrete guidance to data workers on how best to treat data, constitute fundamental starting points towards addressing the thorny issue of data ethics. Most of all, we have shown how regulation and socio-technical systems, while crucial elements in shaping data work, are no substitutes for human judgement and for the capacity to identify and address its broader social implications.

ADDITIONAL READING

Beer, D. (2018). *The Data Gaze: Capitalism, Power and Perception*. London: Sage.

British Academy and Royal Society (2017). Data management and use: Governance in the 21st Century. A joint report of the Royal Society and the British Academy. https://www.thebritishacademy.ac.uk/publications/data-ai-management-use-governance-21st-century/ (Accessed: 15 February 2021).

Couldry, N. and Mejias, U. A. (2019). *The Costs of Connection: How Data Are Colonizing Human Life and Appropriating It for Capitalism*. Stanford, CA: Stanford University Press.

Strasser, B. J. and Edwards, P. (2015). Open Access: Publishing, commerce, and the scientific ethos. Bern: Swiss Science and Innovation Council, SSIC Report 9. https://citizensciences.net/wp-content/plugins/zotpress/lib/request/request.dl.php?api_user_id=424601&dlkey=BDEJXNRZ&content_type=application/pdf (accessed 15 February 2021).

PART IV

DATA VALUE, INNOVATION AND RESPONSIBILITY

Summary

This part examines what makes data valuable, with a focus on the political economy of data. Concepts such as data circulation, platform economy and surveillance capitalism are explained in relation to the value of data. We also analyse why machine learning and the use of algorithms in combination with data lead to the use of metrics and predictions to create value. The next chapter moves the discussion from economic value to ethical concerns about values. We review individual and collective rights, and introduce the concept of data justice. We relate traditional ethical frameworks to data work, and show how ethical issues are present across all interactions with data. Finally, we connect the ethical, economic and political issues reviewed in the preceding chapters to situations where data is used as evidence. The section ends with a demonstration of why it is important to develop responsible data practices.

Learning objectives

This part will help you to:

1. show awareness of the value of data, understand how this value is created, and be able to formulate business opportunities and business models around data flows;
2. consider the role of data as key input and constraint for AI and particularly for machine learning;

3. critically assess the way data shapes science, policy and politics, including how data is turned into metrics that are used for decision making;
4. discuss data justice and fairness in the face of data inequalities;
5. identify ethical issues in real-world projects and know which resources are available to address issues such as data integrity, transparency and security;
6. recognise that governance is a layered process and understand how guidelines, codes of conduct and regulations apply to different data processes and to the use of data as evidence.

Data Story 5: Uber Drivers

Uber is a US-based company founded in 2009 to provide ride-sharing services. In 2020, its activities had expanded to food and freight delivery as well as the development of self-driving vehicle services. At this time, Uber directly employed over 20,000 people, not including its drivers. Its market capitalisation, which is a measure of what a company is worth according to the stock market, was very large, about US$70 billion at the end of 2020. According to Uber, more than 100 million people worldwide used Uber every month.

Uber is most widely known for its platform and apps that support transportation of users by drivers. Using Uber's app, a user can order a ride. The ride request is circulated to nearby drivers. Once a driver accepts the ride request, the driver is expected to pick up the rider at the agreed-upon location and provide a ride to their destination.

- *Is it correct to describe Uber as a 'ride-sharing company'?* Are there other ways of describing Uber? Could we say that Uber is a taxi company, or a data sharing company, or a communication hub? Does it matter how we describe the company's activities? To whom does it matter and why?

Uber is seen as a model for the kinds of 'disruptive' businesses that shape the data economy. Yet, some have argued that its business model is not so innovative: Uber uses technology to displace low-skilled workers in an existing industry, in this case, taxi drivers.

- *Where does the Uber innovation lie?* How does this compare to other data-intensive digital businesses, such as for instance AirBnB (allowing people to rent their living space to strangers) and Turo (facilitating car-share arrangements)?

In many countries around the world, Uber has been the subject of many conflicts, strikes and court cases, with regard to the employment relations its business model supports. A recent vote in California (Proposition 22, November 2020) confirmed that app-based drivers are to be considered independent contractors and not employees.

- *What does it mean for drivers to be seen as contractors?* What can this mean in terms of wages, insurance, responsibility or flexibility for drivers? What can this mean in terms of safety or savings for users?

One of the criteria used in California to classify workers as contractors rather than employees is that the 'the worker is free from the hiring company's control and direction in the performance of work' (California State Legislature, Assembly Bill 5, Chapter 296). A great number of actions on the part of the driver are monitored via the app: rides accepted, rides rejected and rides completed. The app also prompts customers to rate their drivers after the ride. This rating determines whether drivers can continue to work. A driver must achieve a particular score set by Uber (like four out of five stars) in order to be assigned rides. If drivers' ratings fall beneath that score, they are asked to 'improve' by watching a training video that suggests, among other things, that five-star drivers provide a phone charger, bottled water or other items that Uber does not reimburse. Uber also sends its drivers messages via its app about how to behave with riders.

- *What are the implications of Uber standards for drivers?* Are drivers likely to modify their behaviour in order to be on the receiving end of good ratings? Given these interactions between Uber and drivers, to what extent are the drivers 'free from the hiring company's control and direction in the performance of work'?

This approach to using and combining different sources of distributed and automatically generated data is very common in the platform economy. In this case, Uber also uses the ratings of drivers, a kind of reputational data, to organise work. This has been called 'management by algorithm'.

- *How is management by algorithm different from other kinds of management?* Are workers more accountable to their managers or to customers, when reputational data is used? Does this matter?

Imagine a situation where a ride-hailing service driver picks up a group of intoxicated riders, and when members of the group start yelling racist insults at him, he stops the vehicle and asks them to get out, even though the riders' destination has not been reached. The riders give the driver a very poor rating. After contacting the company to explain the situation, the company refuses to remove the poor rating from the driver's profile.

(Continued)

- *What are implications of drivers' dependence on user ratings?* How can Uber distinguish between genuine cases of bad service and cases such as this where it is riders who are at fault? What rights do drivers and riders have within this system, and how are instances of bad behaviour punished? Does it matter that a high proportion of drivers working for Uber are from a minority or immigrant background?
- *What could a driver do to prevent such situations?* Would installing an in-car camera be a useful strategy? Should drivers and riders set up their own surveillance systems to document possible abuse during an Uber ride?

The use of apps and smartphones and the smooth circulation of data are at the core of Uber's service. A smartphone contains hardware and software that generates Global Positioning System (GPS), gyroscope and accelerometer data. When drivers and riders use the Uber app, much data is collected during trips, even though it is not strictly necessary to provide the service (the ride itself).

- *Who owns this data?* Who decides what can be done with the data? When and where are such decisions made?

The data collected is sent to Uber's servers, where it is processed and stored. This data can be used to detect certain driving behaviours, such as rapid acceleration, sudden braking or driving at high speed. Uber explains that analysing this data helps to make their rides safer.

- *How can this data contribute to safer driving?* Who needs to be made aware of this data and how, in order for safety to increase?

This data is also used by Uber to help develop self-driving vehicles. By using the data it has collected across millions of rides, Uber can model how self-driving cars may interact with specific environments. The analysis of this data might also help identify geographical areas where self-driving vehicles can best be used, or the kinds of rides that can best be performed by a self-driving vehicle.

- *What are the implication of planning future transport on the basis of Uber data?* The data is collected by Uber depending on demand from riders. How does this shape the data? Might it emphasise particular kinds of routes and ignore others? Are certain types of rides at certain times of day overrepresented in the dataset? What could this mean for the development of self-driving vehicles? Does this use of data to automate transport help the drivers who helped create the data?

The Uber app also coordinates where and when drivers work through economic incentives. The system does this by nudging drivers. It announces higher rates in order to attract workers to work in areas or at times of the day that are expected to have increased demand. These predictions are generated by algorithms that are fed with the data collected by Uber.

- *Who benefits from demand and offer being shaped by the platform?* Is this price fixing and does this matter? To what extent can drivers understand and control how their earnings are being set?

In 2019, Uber was responsible for a significant proportion of rides/car movements in North American cities. Uber also started to develop services for food delivery and logistics (parcel delivery), which can be seen as an extension of their ride-sharing business. Uber also worked to connect ride-sharing to public transportation through collaboration with city councils. Typically, these collaborations aim to provide specific transportation services that connect the use of private cars and buses, trains and metros; for example, to link homes and transit stations, or late at night when public transportation services are running less frequently or not at all. These collaborations take different forms: from offering vouchers for a discount when Uber is used to travel from a train station, to subsidised rides for passengers with a disability.

- Can reliance on Uber help reduce congestion and parking problems in cities? Could it result in passengers moving away from using buses and trams?

Most public transport services are subsidised. This is therefore an example of public and for-profit activities becoming entwined.

- If apps start to connect public transit and Uber services, which novel data does this allow Uber to collect? What are the implication of a public service becoming dependent on corporations in such arrangements?

Data Story based on:

Eyert, F., Irgmaier, F. and Ulbricht, L. (2020). Extending the framework of algorithmic regulation. The Uber case. *Regulation & Governance*. https://doi.org/10.1111/rego.12371

Mosseri, S. (2020). Being watched and being seen: Negotiating visibility in the NYC ride-hail circuit. *New Media & Society*, October, 1461444820966752. https://doi.org/10.1177/1461444820966752

Rosenblat, A. (2016). The truth about how Uber's app manages drivers. *Harvard Business Review*, 6 April. https://hbr.org/2016/04/the-truth-about-how-ubers-app-manages-drivers (accessed 3 May 2021).

Data Story 6: Dating Apps

The use of 'dating applications' to find companionship is well established around the world, with millions of individuals feeding copious amounts of personal information to these programs in the hope that they will help match them with compatible people, and eventually help them to find a partner or other forms of companionship. This new way of looking for love, sex and friendship has had a lot of success. It is not difficult to see why: dating apps extend our social circles well beyond people known to us or to close friends, and thus increase the chance of finding a compatible mate or sexual partner. Furthermore, the opportunity to personalise these services makes it possible to look for people similar to ourselves – or/and mirroring the qualities we look for in a partner. This of course means that the more personal information we insert in dating apps (including photographs of ourselves, favourite hobbies and venues, family and educational history, preferences around sex, sports, food and arts), the higher the chance to be matched with compatible people.

- *Is all the information we insert into dating apps 'data'?* After all, they were never generated with the intention of creating a body of knowledge; rather, they are used to support communication among the app users. In which sense are these bits of information data? For whom? Under which circumstances?

Some of the people who use dating services are not very concerned about providing personal information. In their view, it is no big secret that they like tacos and enjoy going to the gym, and they do not think that anything bad could happen to them should this information become public. They feel that they have nothing to hide, and therefore do not see issues with data security in relation to the use of dating apps.

- *Are all such data sensitive?* And are they all 'private' data? What is the difference between 'personal' and 'private' data? What uses can you imagine for such data, and by whom?

Other people are concerned about the data they insert into dating apps, and particularly about whether there are any mechanisms to check whether what people say about themselves is true. These apps have notoriously been used by criminals to approach vulnerable victims; for instance, by luring individuals into secluded spots in order to attack them, or by establishing an emotionally binding relationship with someone in order to steal money from them (grooming). Less threatening but still worrisome is the chance that people will unduly enhance their profiles;

for instance, by using pictures of themselves that are digitally modified, from a different period of their life (in order to look younger, for example) or even pictures of different people altogether. Using such inaccurate pictures may feel like a betrayal for the prospective date once the individuals in question meet in person.

- *How do dating apps generate trust?* Is it right for people to trust the information found on these sites? What kinds of verification services do they rely on? Should users be aware of whether and how information is verified? Should they worry about it?

The information from these apps can also be valuable for other purposes than arranging dates.

- *What possible reuses of these data are acceptable in your view, and why?* Is it acceptable for public health researchers to use these data to develop measures of loneliness and mental health in a population? Is it acceptable for a company to use these data to develop marketing strategies to sell singles holidays, plastic surgery or self-improvement courses?

By their very nature, dating apps encourage users to disclose information about their dreams, expectations and preferences. Whether or not a user thinks of such information as private or even intimate, disclosing it makes the individual vulnerable in some respects. There are dating apps that actively tackle this problem by giving more control to whom they view as the more vulnerable users. For instance, some dating apps give women the power to make the first move by picking the contacts they are most interested in, while men are given the more passive role of accepting or rejecting women's approaches.

- *Do dating apps have a duty to protect their users?* Are some groups of users (such as young women, for instance) more vulnerable than others, and thus in need of higher protection? Does allowing women to 'make the first move' redress power dynamics or does it reinforce stereotypes? To which extent is this culturally sensitive, and should dating apps have different settings depending on where they are used?

8

DATA AS A SOURCE OF VALUE

Summary

New types of data are created in the process of datafication, often making everyday, mundane activities amenable to all kinds of investigation or calculation and positioning them as sources of value. New types of services are also changing how value is generated. These new data practices create value in a diversity of ways, though there are also dominant configurations of actors, technologies and practices. In this chapter, we sketch the concept of the data economy and ask how data is valued as a source of innovation. We explain its role as a source of economic value, and analyse the regulation of various market relations in which data is involved. We also describe how the production, processing and combination of data as well as various AI applications lead to valuing data in specific ways.

8.1 Introduction: What makes data valuable?

To answer this question, we have to define value. Value can be the amount of money paid for a good or service. Value can also be the affective importance or practical usefulness of something or someone. Value can have a moral sense and refer to how one aims to live and distinguish right from wrong. Depending on what we mean by value, different systems and mechanisms come into focus. If we speak of economic value, then we need to consider capitalism, market dynamics and the regulation of transactions.

Yet, like many other aspects of data discussed in this book so far, these different forms of value are entwined and can be simultaneously present or follow from each other. For instance, data have political and financial value, as the result of costly investments, as tools to legitimise or oppose governmental policies, or as trade currency among national governments, lobby groups, social movements and industries. They have cultural and social value through their link to the identity, histories, norms, sensitivities and behaviours of the communities and individuals that they are taken to document. And data can have affective value for those who invest time and effort in their production and interpretation, as well as for the human subjects from whom data are extracted. What becomes clear when considering the diverse forms of value attributed to data throughout their journeys is that it is this multiplicity of value that makes data into useful sources of evidence.

The value attributed to data each time they are used has a defining effect on the way data are managed and interpreted. Furthermore, the entanglement of different forms of value attributed to the same data is key to their economic and commercial value (Tempini, 2017). Consider again personal data extracted from social media like Twitter and used to produce medical knowledge, as discussed in Chapter 7. These data have affective and cultural value, insofar as they are fundamental to the feeling of identity of individuals and, when shared, provide a foundation for the development and resilience of social groups (Harris et al., 2016; Leonelli, 2016a). They also have scientific value as a potential sample of social behaviour, which can be studied to extract generalisable knowledge about human relationships, socialisation and opinions on a variety of subjects. Far from constituting a conflict, this intersection of affective, cultural and scientific value is ultimately what makes these data commercially valuable.

Besides the question 'what kind of value' or 'whose values', another important debate about the value of data concerns whether some types of data should even be valued in economic terms. We have seen in Chapter 7 how data can be considered by different actors or in different settings as either proprietary or as a good that should not be commodified – a distinction that also relates to our introductory discussion of the plural and singular form of the word data. For example, when it comes to personal data, some argue that the current problem is that those whose personal data is used to generate profits are not properly compensated. To remedy this unfair situation, new regulations about property rights of data should be put in place. Others argue that personal data should never be seen as a commodity and that data governance should aim to maintain such data – if used at all – as a public good. We note this issue for now and return to it in the section on regulation that follows.

We will focus on the economic value of data. In a capitalist society, economic value plays a central social role: it connects markets and governance, and shapes our actions and opportunities as consumers and citizens. At the same time, any discussion of value is not only about profits and shareholders but also about accountability, responsibility and freedoms. Current discussions about the value of data often touch on the kinds of societies one wants to achieve and on the opportunities one wants to foster for citizens. The Organisation for Economic Co-operation and Development (OECD) for example makes a strong connection between data, innovation and a thriving society, insisting on 'the use of data and analytics to innovate for growth and well-being' (OECD, 2014). Similarly, the European Union (EU) connects data with economic and social welfare, as well as security, in the concept of technological sovereignty. If the EU's future is tied to data and related technologies, then EU policy needs to achieve 'technological sovereignty in certain critical areas such as blockchain, high performance computing, quantum computing, algorithms and tools for sharing and using data' (Von der Leyen, 2019). In these terms, the value of data is also a political and societal concern. Finally, while the emphasis in this chapter is on the political economy of data, it remains important to realise also that data can be a resource for other purposes than control and money-making. For example, an in-depth study of self-trackers and how they use self-tracking technologies for a range of practices, including mindfulness or self-care, shows that data can have a very different kind of value (Sharon and Zandbergen, 2017).

8.2 Assumptions about the value of data

In discussions of the value of data, we often encounter a number of assumptions. As we saw in the previous chapter, data is often defined as a non-rival good. While some scholars have fruitfully analysed data as capital rather than as a good (e.g. Sadowski, 2019), we prefer to treat it as a good or asset to keep the materiality of data firmly in view. This is the assumption that use of data by one party does not reduce its availability to others. This can be illustrated by comparing it to a rival good: If I eat an apple, it is no longer available as food for others. In contrast, if I read an e-book, others can read it too. What we want to stress here is that the non-rival feature of data is often overemphasised: data also has a material aspect and has to fit in our practices. There are therefore limits to this non-rival nature of data. Accessing an e-book requires other resources that are clearly finite (technological devices, infrastructural bandwidth, energy). Furthermore, there is only so much data one can pay attention to, so many e-books one can read in a day – selection mechanisms are also at play. Data is not immaterial and its availability does have limits. The production and use of data requires bandwidth that depends on physical infrastructures, servers require energy to run and be kept cool, devices must be produced and there is also a limit to human attention that is not endlessly expendable.

Another assumption is that data are generic and have no value in and of themselves, so that value creation around data becomes the key challenge. This assumption is used to drive data sharing and data circulation, since they are put forth as the main or even the only route to value creation with data. In this view, any barrier to data circulation is an obstacle to innovation and value creation. Often, the only acceptable limitations seem to be those that maintain some forms of privacy. One effect of this is to undervalue some kinds of data and to position privacy in opposition to value creation. As a result, privacy and innovation come to be placed in a false dichotomy. This opposition has problematic implications for data curation, since attention to data privacy and confidentiality concerns may actually enhance the value of data. If better privacy leads to the development of data infrastructures aimed to enhance data reuse, then attention to privacy actually increases data use and circulation. Another consequence of undervaluing privacy is that, arguably, users whose data is collected are not adequately compensated. A classic manifestation of this is the provision of cost-free email accounts by corporations, in return for access to all data generated by the user when making use of this email account. The value of the service provided is not proportional to the value of the data created, and only those able to pay for an email service are able to avoid this unfair trade-off. Privacy thus becomes a luxury good rather than a fundamental right.

Finally, there is the problematic expectation, as discussed in the previous chapter, that current regulations and bodies can ensure proper governance of data activities. For example, contract law is expected to regulate and fully control relationship between corporations and users. Yet there has been ample demonstration of shortcomings; for example, the impossibility of users actually 'reading' the terms and conditions even for a single app (which can amount to 250,000 words; Kaldestad, 2016) and thereby providing informed consent. As we saw in the previous chapter, governance is a layered process. All these assumptions and expectations around value shape how we assess what we can and cannot do with data.

8.3 Data and innovation

In the rest of this chapter, we examine how data comes to be valuable and review the contexts in which these assumptions may or may not hold. New types of data and data linkages can help constitute a new source of knowledge and lead to innovative approaches in policy making, research or marketing. To begin, we will look at how data is positioned as a source of innovation. Data can be valuable for improving the functioning of organisations, creating new business models and labour relations, and revealing hidden value through pattern recognition and prediction.

8.3.1 Improving organisations

Some organisations have an identity that is strongly linked with data. Think of the label 'BigTech', BATX, for the largest internet corporations in China: Baidu, Alibaba, Tencent and Xiaomi, or FANGAM to refer to large corporations such as Facebook, Apple, Netflix, Google, Amazon, whose business models are built around the collection and analysis of user data. Even beyond these well-known corporations, the majority of organisations, whether public or private and across all domains of social life, increasingly rely on data to monitor, manage and develop their activities. They therefore all share some aspects typically associated with tech companies, including the focus on data collection as a source of evidence for how to act. This trend did not begin with Big Data but is part of a longer history of aspiration towards the 'automation' or 'digitisation' of society and related services. The use of Big Data is often associated with increasing efficiency in organisations, in order to reduce costs of operations, to better serve customers and to make organisations more competitive (Kitchin, 2014). Many companies rely on data to run their operations; for example, online shops, warehouses and deliveries are tightly interlinked through data. The use of data in organisations is often predicated on the hope of making organisations more efficient and more integrated.

8.3.2 Generating new products, new business and new kinds of work

There are new types of activities that have developed around data, including selling data products. There is no universal pyramid of value attached to data at different stages of data work – all may be regarded as valuable depending on who is looking, why and how. Unprocessed data (data that have not yet been cleaned) are often traded as valuable commodities, even if they are more difficult to use than processed data. At the same time, there is also a whole industry around data enrichment, which typically consists of taking a given data product as a starting point for a new usage of those data, thus providing added value and building on the initial promise of data to deliver meaningful insights. For example, consider genomic data. The string of sequencing data in itself is not meaningful, despite the enormous amount of effort, data cleaning/ modelling and related investments required to create that particular data product. The 'real value' of sequencing data lies in their subsequent enrichment; for example, when coupling them with information about gene expression (to understand how a particular genetic trait ends up affecting a particular part of an organism) and phenotypic traits correlated to the existence of that genetic trait (to understand for example whether a given symptom of cancer is associated with having a given trait). In biomedicine, a large industry comprising both big pharma and myriad start-ups is devoted to building data enrichment tools and related diagnostic instruments.

Platforms are especially important for creating value through the use of data. Platforms achieve this through their role as intermediaries. They connect different players in the market through digital tools that shape how these players can interact with each other. Platforms rely on the internet and the penetration of information technology, which make interactions digital. Further datafication meant that these interactions also become traceable and computable. As cloud infrastructures were rolled out, platforms aligned to these services and became even more dominant, to the point of restructuring entire aspects of the economy. A number of labels have been used to describe these dynamics: platform economy, creative economy or sharing economy. While platforms such as GitHub, Napster, Wikipedia, Zipcar, or that which produced Linux, might be labelled sharing economy, many other have forms that are much more oriented towards monetising interactions and extracting data and labour. These have been associated with platforms such as Uber, Mechanical Turk (where small tasks are set out for workers to perform online for a small financial reward) or AirBnB and labelled the 'gig economy' and associated with new forms of capitalism, and often less positively, with the 'digital precariat'.

These ways of making a living are heavily reliant on data. The gig economy as a new way of valuing labour (food deliveries and all kinds of tasks ranging from carpentry to writing reviews to care work) is also highly reliant on the flow of data about demand and supply. Vloggers, influencers and mommybloggers all monetise the value of their work using data from likes, followers and clicks. Career opportunities can also be focused on data: hackathons and data competitions are ways to demonstrate the value of one's work to potential employers. These developments are often claimed to enable more people to pursue creative or less conventional learning and career paths, and to act as entrepreneurs. There is however a tendency for most of the control of wages and work to be with large organisations, while risks and precariousness rest with workers.

8.3.3 Finding hidden value

Finally, data is part of an approach that seeks to find hidden value by identifying patterns and making predictions. A well-known example of this was popularised in the film *Moneyball*. In this narrative, an underdog, underfunded baseball club uses data analysis to forecast the potential value of players and thus uncover their hidden value as future athletes. This provides this club with a competitive advantage, since it is able to acquire undervalued players cheaply and to set them to work in a way that makes best use of their talents, as revealed by statistics. This is a classic alignment of data–discovery–transaction–management that has a looping, self-reinforcing effect (discussed in the section on 'The Value of Metrics'). While this example is from the world of professional sports, this approach is also at the core of many applications in science and research, and data analytics companies have developed precisely around this particular kind of data value.

Clearly, these three ways in which data is positioned as a source of innovation often intersect: Amazon is both a retail platform and a 'gig' platform via Mechanical Turk, as well as a pioneer in data analytics (older readers will recall the novelty of being told that 'customers who bought this book are also interested in these other books'). In the rest of the chapter, we focus on how value is created with data and by whom, under what conditions. These considerations help us understand the political economy of data.

8.4 The data economy

So far, we have considered data as a source of innovation. Before going further, it is important to note that in spite of the spirit of invention and even of revolution that often accompanies discussions of the data era, many historical elements of inequality or exploitation have remained or even been amplified in the data economy. This can take the form of collecting information in especially intensive or invasive ways on particular groups that are poor, or migrant, in need of help or otherwise disenfranchised (Eubanks, 2018; Couldry and Mejias, 2019). Also deeply problematic are the growing precarity of labour, as platforms have come to articulate demand and supply, and the strategic use by BigTech of tax havens and other ways of shirking fiscal responsibilities. We will discuss in more detail ways of addressing these issues in Chapter 9.

In this section we will consider what shapes the current 'market' and data relations. It is especially important to understand how data comes to be seen as a good, how network effects increase the value of data and increase the hold of monopolistic actors on data. One key dynamic is that a new kind of vertical integration is developing, where firms take advantage of the opportunity of 'capturing' users through managing a whole system of hardware, software and interfaces/services. Together, these elements form the data economy, where control over data and the promise of data as investment and capital are especially important.

The processes that make data a valuable good are complex and rapidly changing. This is visible in the way the public sector increasingly relies on companies that provide data services, so that companies become an essential part of service infrastructures. A recent example of this is the proliferation of commercially produced contact tracing apps in relation to COVID-19 that displace other activities and come to form an indispensable part of public health practices (Aouragh et al., 2020). Another important dimension is the appropriation and exploitation of resources in ways that are arguably not compensated or reciprocated fairly. In the common case where data from social media is used to produce valuable insights for companies, the creators of the data may not even be aware that their data labour is yielding something commercially important. In many cases, there is a barter going on, where users agree to give up any rights to their data in return for the provision of a service (e.g. a free email service, social media interactions, watching videos on YouTube). The key question here is whether the value of the data given by the users is proportional to the value of the services provided. If we look at the profits of BigTech, it seems that quite a lot of value is being created. But to be able to judge such issues, we first need to look more closely at how data becomes valuable.

At present, platforms are important in producing and shaping data and in making it valuable. We discussed platforms in Chapter 6 as an important aspect of the technological architecture of the digital sphere. Platforms are also a business model that enable one to gain a competitive advantage and to create new forms of value (Srnicek, 2016; Van Dijck et al., 2018). Platforms, insofar as they have become a prime site for extracting, processing and analysing data, have also become a site to generate profits and to compete in the digital economy. Platforms tend to be monopolistic due to network effects. This is the case when the value of using a particular service offered by a platform increases because of the number of users already using the platform ('everybody is on Instagram'). This is also what makes it hard for users to leave a platform. An interesting strategy to counter this difficulty is adversarial interoperability: creating a new product or service that connects into the existing ones without the permission of the companies that make them. This strategy can ease the shift from one service provider to another and decreases the dominance of the larger players.

Nearly all platforms are commercially owned. It may even seem difficult to imagine a government-operated platform, yet the internet was originally entirely publicly funded and closely embedded in public institutions. Currently, companies that develop and operate platforms position themselves as intermediaries between different groups and tend to highlight their role as making interaction possible – for example, a food delivery platform mediates the interaction between hungry customers and restaurants offering food. Their 'business' is the provision of information about customers to restaurants and information about restaurants to consumers. At this level, platforms therefore combine a technical and a business vision: their design makes it possible for new informational services to be providers for users, for which the platform can ask for a small fee (typically a commission per transaction; the restaurant pays a given amount to the platform for each order placed).

There is another layer of activity that is very important for the creation of value via platforms: the data gathered from the *use* of these services is also valuable. To understand this layer, we need to think back to our discussion of the dynamics of datafication, where everyday activities – as well as potentially all interaction with digital technology – produce 'traces' that can be gathered. These traces stand in for all types of relations and objects: 'friends' on Facebook for social relations, 'likes' for preferences, use of certain words on Twitter for 'sentiment', etc. Whatever we may think of the validity of the assumptions made about what these traces stand for, and the variation in their meaning ('like' or 'friend' can mean different things), platforms are an important site where these traces are generated and gathered.

As noted above, platforms owners tend to present their business as being the provision of information, for which they collect a fee. Uber presents its business as connecting drivers and riders, and not as providing a transport service. As such, Uber places itself as being outside the transport sector and therefore not subject to the regulation of that sector. This positioning has been at the centre of a number of court cases, where the point of contention was whether platforms do more than provide information and therefore should bear the responsibilities of roles that extend beyond that of information provider. This positioning has implications for value, since information exchanged is only part of the value creation that is occurring on platforms. Besides providing information to its users and matching supply and demand, platform owners are also gathering data from the behaviour of all of its users as they pursue the activities in question (requesting and providing rides, deliveries, etc.) and as they use the platform environment. In other words, at the same time that platforms gain from each transaction on the platform, they also extract data about these transactions as well as how users behaved before and after making these transactions, in a move you could label 'dual value production' (Van Doorn and Badger, 2020). This additional data is potentially very valuable.

The creation of value from data on platforms generally includes the following steps: platforms engage users in the creation of traces (liking, friending, browsing, ranking, etc.); users are construed as aggregates of actions based on these traces; and users can be compared on the basis of patterns in these traces. These three steps can be labelled encoding (the production of traces), aggregation (the creation of a group or classification) and computation (the use of data analytics techniques (often algorithms) to find patterns; Alaimo and Kallinikos, 2017). In the early 1990s, Phil Agre proposed a similar analysis to analyse what happens to human behaviour when it becomes digitised in his 'capture model' of surveillance (Van Doorn and Badger, 2020). He discussed how close analysis of human behaviour, followed by standardisation and articulation of labels to describe the various components of behaviour, could lead to imposition and instrumentation of particular forms for activities. Agre's concern was in line with the dominant concerns of the 1990s, regarding how work was being automated and jobs were disappearing. Since the formulation of Agre's model, one can speak of datafication. A number of additional developments have occurred since that amplify the effects of such capture surveillance:

- Coupling of kinds of data, across activities: the more platforms are integrated with all kinds of other technologies (like Alexa, Google Nest, sport watches, smart washing machines), the more data is collected and aggregated. These can in turn form the basis of services for users but also for advertisers, political organisations or government.
- Scaling up: the potential of this data depends in part on the scale at which it can be collected. In order for patterns to emerge, large amounts of data are needed.
- (Infra) structural combination of data and analytics: part of the value of this data depends on the possibility of linking it to the algorithms being developed by platform companies. Algorithms become more valuable as they are trained on additional data, while data becomes meaningful when analysed with well-functioning software.

8.5 Who benefits from the value of data?

The combination of access to data generation, the possibility of doing this on a large scale, and the control of both data and algorithms are elements that put platforms in a very strong position when it comes to monetising the value of data. 'Data analytics' is a way of revealing patterns that would otherwise remain hidden and this is best done through the use of quantified data, systematically obtained from collection tools and platforms. The data and the conclusions drawn based on it are in turn affected by the activities that produce the data – a feedback loop that we discussed in Chapter 3.

As such, the data is not so much emergent and revelatory: data analytics re/produces a world in which data analytics matters and therefore further establishes grounds for its own value (feedback; looping effect). For instance, the value of Facebook in providing data on consumer preferences and behaviour depends on the frequency with which its users interact with Facebook and make use of its advertisement services. More user interactions mean more data, which in turn makes Facebook advertising more effective and better targeted to users. In this way, Facebook data analytics reproduces and expands its value over time. Another example is that of citizen science initiatives such as CurieuzeNeuzen, a Belgian project that distributed tools to measure air pollution to 20,000 citizens and used the incoming data to assess and monitor air quality. This project was so successful that similar initiatives sprang up in neighbouring countries and managed to acquire funding, thus creating an ever-expanding system for the collection and analysis of air quality data, relying on a growing group of participants and analysts (Irwin, 2018).

As these examples show, platforms not only have an economic effect, but also shape labour relations and the organisation of aspects of public life. These developments can have considerable beneficial effects for public engagement in political, scientific and social projects – for instance, when the sharing of health data across countries helps to diagnose and treat rare diseases. Another example would be to provide tools to record and highlight human right infringements (such as capturing and sharing instances of violence via Twitter). This data can help to identify and sanction instances in which governments or law enforcement bodies such as the police abuse their citizens. Platforms also have consequences for gendered and racialised types of work while enhancing the accumulation of wealth for a few. While there is much evidence that the platform economy tends to increase disparities (Drahokoupil and Jepsen, 2017; Hoang et al., 2020), there is no single effect of the valuation of data. For some countries in the Global South, flexible and formalised work associated with the platform economy can create opportunities as well as contribute to a reduction of wages (Abrieu et al., 2019). At the very same time, the use of platforms is also closely tied to the proliferation of neoliberal forms of capitalism (Beer, 2018; Mirowski, 2018; Zuboff, 2019). Neoliberalism favours deregulation, privatisation and reducing state influence on the market. This has created the conditions that favour large global players and monopolies, that reduce the limitations on what can be monetised and that decrease regulations that hold corporations accountable.

Many of the examples discussed so far fall squarely within the critique of the data economy that highlights its extractive and exploitative aspects. Many worry that BigTech companies increasingly take on the role of public actors and that we are becoming dependent on them for social interaction, public debate or provision of public services. Given the dynamics of data extraction of platforms described above, all these activities potentially become part of money-making models and practices. While states and public bodies have been a key part of how we organise society, the concern is that we will not be able to sustain state activities and services without BigTech (Magalhães and

Couldry, 2020). While utilities such as water and electricity providers are heavily regulated and the extent to which they can make a profit is limited, this is hardly the case for platforms. But if we become dependent on platforms in the same way that we depend on electricity and water supplies, this raises the question of whether these actors should be more highly regulated and accountable (*The Economist*, 2020).

The current arrangements have allowed the proliferation of platforms in the hands of a few dominant actors, but these are not necessarily set in stone. Potentially, other actors can also have access to data. For example, General Data Protection Regulation (GDPR) provides the basis for one to obtain one's data from a given platform. However, without access to the software/analytic tools on the one hand and to data on a large enough scale, it is hardly feasible to valorise such data or to produce analyses of a sufficient scope to counter those of the big actors. This has led to the suggestion from different corners that we need platforms that function as public utilities or that foreground public values in order to connect data and data analytics in a non-commercialised setting (Srnicek, 2016; Van Dijck, 2020). Such calls have been especially prominent in the wake of the 2020 pandemic, when many realised our dependence on corporate-owned infrastructures for pursuing social, educational and political activities. These alternatives may seem difficult to imagine. To illustrate such possibilities, one example is that of the Workers Info Exchange (WIEx). It brings together workers (from Uber), academics, legal experts and computer scientists to build an infrastructure to which data can freely be submitted to create a larger data pool that can be used for the benefit of workers. The website enables workers to mandate WIEx to obtain their data from platforms. The platform also links experts who are concerned about practices of robo-firing or firing by algorithm, where Uber drivers have been dismissed due to algorithmically detected fraud without the provision of evidence or the possibility of appeal. Such forms of defence of worker rights, via data collection and data analytics, are a response to changing labour relations. It remains a considerable challenge to act on the scale necessary to provide alternatives to platforms operated by BigTech. For example, the longevity of alternative public tools is difficult to maintain. Software needs both frequent updating and long-term user support after the initial development phase. This is especially challenging for smaller organisations, whereas BigTech profit from constant updates (Chun, 2016). We will consider more examples of responsible data practices in Chapter 10.

8.6 Allocating value, responsibility and profit

Our discussion of value so far shows that it connects data activities to markets. There are various mechanisms that regulate the data economy. For example, there are anti-trust laws that aim to maintain competition between firms and ensure that there is a level playing field for all actors. These aim to prevent price fixing and monopolies. Most of the BigTech companies, including Facebook and Google, have been the object of anti-trust investigations in recent years. These laws are usually enforced by imposing fines, which may not be significant to large corporations.

Another familiar mode of regulation is the establishment of a contract between users and service providers. In the digital world, we encounter such contracts frequently – for example, when we want to download a new app – in the form of end-user licensing agreements (EULA). In these contracts, users are provided with the terms and conditions of use and agree to allow the service provider to use their data in a number of ways. Such EULAs have been amply analysed and shown to constitute a legal but tenuous kind of agreement: these contracts are effectively and practically impossible to read because of the complex legal language, length and frequent updates. Such contracts can be said to constitute acquiescence at best, rather than informed consent.

Labour is also widely regulated, though different political cultures put varying emphasis on collective protection versus individual freedom and responsibility of workers. The reorganisation of work in the 'gig economy', where platforms shape labour relations, has caused many forms of actions, including, strikes, lawsuits and attempts to reform labour regulations to address these new conditions of work. One of the key problems is the relationships between workers, platform owners and employers, as discussed above in relation to platforms. When there is a relationship of employment, certain obligations are attributed to the different actors. When platform actors position themselves as simply information providers, this sets them outside of such an employment relationship. The result is that the platform actors do not have to act as a responsible employer, even though the platform is creating the conditions of employment (in the case of Uber: deciding how drivers should behave, allowing or denying them access to the platform, etc.). Several court cases have been fought with very mixed results about how the different roles are defined. The nature of the relationship between workers and platforms also has fiscal consequences. For example, depending on whether Uber drivers are independent contractors or employees of Uber, who is responsible for administering taxes and other insurance obligations (social security, unemployment insurance) changes. This means that a company like Uber can face more or less administrative costs depending on how their role is defined.

Finally, how proprietary rights are assigned to data also constitutes a highly dynamic area. The multiple forms of value attached to the data can give rise to tensions and disagreements among potential users, and between data producers and data users, as discussed in Chapter 7. Data has also become an object of global policy making and international regulation. It is significant for economic relations and security. For example, the recent EU policy document highlights the importance of data and data sharing. This brings us back to issues of data flows: if international policy shapes access to markets, prevents export or requires licensing (often extra-territorial – to use an American product in other countries is subject to US-formulated licences), this has consequences for how data can be shared. An additional concern is that many data-intensive technologies have both civilian and military applications, thereby raising security issues.

8.7 How does AI add value to data?

Much attention is currently paid to issues that arise when data is used in the context of algorithms, machine learning and AI. In the final part of this chapter, we consider how the combination of data and AI raises issues with regard to the value of data. Automation of data processing and the use of machine learning make it increasingly easy to extract certain kinds of value out of data (e.g. better predictions, less human labour involved, engagement with data production through feedback). At the same time, this intersection of data and AI also aggravates issues that are already present, around predictions and metrics.

We have discussed how data from platforms becomes commensurable, in the sense that it can be easily combined and compared. Increasingly, the possibilities for combining data are built into the tools and platforms used. Data mining and machine learning are used to perform data filtering and classification, to increase efficiency, and to address optimisation and strategic recommendations. In these settings, flows of data become valuable for the ways they can be used to make correlations and contribute to predictions. This emphasis on the value of data systems for forecasting can sometimes lead to underestimating or even disregarding questions around the quality of the data, the plausibility of the analysis and its broader social implications.

One of the key expectations attached to the previously discussed forms of data analysis is that they will become increasingly automated. Such automation of analysis can be labelled **artificial intelligence (AI)**. AI has a history that is at least decades long and does not begin with Siri or self-driving cars. It is also a mix of many disciplines (see Chapter 6) and combines statistics, modelling, programming and computing, and handles many different forms – numbers, words, images, sensor data. AI has been widely used in engineering and cognitive science contexts, and in the past decade has been especially intensively extended to many other areas, from the mundane to the very sophisticated.

In particular, one branch of AI called **machine learning** has widely expanded and proved very useful in certain areas. Machine learning is the process by which a dataset is used by a computer to build and/or further refine an approach to solving a specific problem, such as image recognition or classifying information. Machine learning starts with creating a model in a language a computer can process (frequently used are R and Python), and then training this model on a dataset. These models draw heavily on statistics and on probability theory. In this section, we focus on how machine learning plays a role in the valuing of data.

Machine learning involves automation and calculation. While all modes of calculation involve abstraction, machine learning pushes this to a greater extent. Typically, within machine learning applications data are ordered in tables and given a specific weight, which is then calibrated and modified as the algorithm is trained on specific data samples (often called 'reference data'). The focus is on exploring the relationships within and between datasets that have been brought into a common space of calculation. While machine learning has its own varied history going back several decades – like so many other lines of work discussed in this book – machine learning currently focuses on finding a mathematical function that could have generated the data, and on optimising the search for that function as much as possible (Mackenzie, 2017). Deep learning is a particular category of machine learning, associated with the ground-breaking work of computer scientist Rina Dechter. The 'deep' part of this machine learning refers to its layered or nested structure. At each layer, data is transformed into more abstract representations that feed the next layer. This is an iterative process where the structure set up by programmers is quickly superseded by updates 'learned' and implemented by the algorithms themselves, as a result of training on new data. The results are often inscrutable to humans, with ever more complex algorithms yielding outcomes that cannot fully be explained even by those who have developed the technology in the first place. The outcomes of machine learning are often 'classificatory statements' (this is a picture of cat) or 'predictive statements' (we will need 8% more drivers in the city centre on a rainy, late afternoon that falls on a Thursday).

An often-heard term associated with machine learning is '**algorithm**'. This label refers to the operations and calculations that machine learning performs on data. Algorithms were first formulated in the field of mathematics. They are a way of formalising an activity, by specifying a set of steps that must be followed. Because computers have been developed according to a formal logic, algorithms are a good fit with how they work. In machine learning, computers are used to build models that are based on training data. The outcome is usually a prediction or a classification.

Typically, algorithms are evaluated on the criteria of correctness (do they provide the right answer) and efficiency (in the computational sense, often translated as how long does it take to get an answer). Often, there is a trade-off;

the aim is to get a good enough answer in a reasonable amount of time. Of course, depending on context, what is good enough and what is a reasonable amount of time will vary, whether the aim is to find a good route to drive to a particular holiday destination, a suggestion for which TV series to watch, the oxygen needs of the International Space Station or the likelihood of a tumour being malignant. Developing good algorithms requires the combination of multiple types of expertise (as discussed in Chapter 6), where statistics and mathematics need to be complemented by expertise in programming and computer engineering. A problem like overfitting illustrates this. Overfitting is the mistaken identification of patterns in a dataset, which can be greatly amplified by the training techniques employed by machine learning algorithms. An algorithm that is trained to extrapolate patterns from a given dataset may not be as successful when applied to other data. A common approach to address overfitting is to divide up the data and compare how the algorithm performs on two subsets of data (cross-validation) or to compare how two differently trained algorithms perform ('ensembling'). The point is that this kind of work is data-centric (Leonelli, 2016a). The procedures, techniques, methods, software and hardware are determinant for the insights that are produced using algorithms. This means that statistical logic or mathematical coherence cannot be applied as the sole criteria to evaluate an algorithm. The combination of data production, analytic methods, statistical model, computational aspects and hardware all contribute to shaping algorithms. (This is partly why they are so difficult to evaluate, as we will discuss in the next section of this book.)

The reason there is so much talk of algorithms and machine learning is that they have been incorporated into all kinds of practices that we associate with datafication. Yeung has labelled this algorithmic regulation, defined as:

> decision-making systems that regulate a domain of activity in order to manage risk or alter behavior through continual computational generation of knowledge from data emitted and directly collected (in real time on a continuous basis) from numerous dynamic components pertaining to the regulated environments in order to identify and, if necessary, automatically refine (or prompt refinement of) the system's operations to attain a prespecified goal (Yeung, 2018, p.507, quoted in Eyert et al., 2020).

This conceptual description of algorithmic regulation corresponds to what we have discussed in the Data Story on Uber drivers. The decision-making system of Uber regulates the assignment of rides to drivers, using data emitted and collected from drivers, traffic and ratings of riders among other data sources. This data is processed by an algorithm that classifies and prioritises different aspects. The knowledge of which aspects are used and precisely how they are used constitutes very restricted knowledge, as this algorithm is part of Uber's business intelligence. The algorithms assign rides so as to set out the work as efficiently as possible, ensuring the maximum number of rides being completed to the greatest satisfaction of riders, and thereby growing the market share of Uber. We have also seen that drivers try to figure out how the algorithm works to be able to act in ways that benefit their own functioning in the ride-sharing business. But the algorithm and its workings are proprietary and belong to Uber, which does not involve other parties in its development and is only accountable to its shareholders. Uber is but one example; there are many other cases that affect health, education, employment and access to benefits or even freedom, if we think of the case of parole from prison also being shaped by algorithms. This means that algorithmic rationality is increasingly important for how particular kinds of services are organised, for how we make decisions and how we act. However, because algorithms are ubiquitous and opaque, their use can reinforce particular tendencies and leave us little recourse. We discuss this further in the next section.

8.8 The value of prediction

Algorithms enable software to analyse data and make predictions about outcomes. These can be highly accurate. Among the factors that determine this accuracy are the quality and volume of reference data. This is yet another way in which data can be valuable. In turn, the ability to produce accurate predictions makes some algorithms very valuable. To return to the Uber case, the Uber algorithms aim to predict demand and ensure sufficient availability of drivers to meet this demand. In order to make such predictions, access to data to build, feed and continually improve algorithms is essential. The label 'surveillance capitalism', articulated by Shoshanna Zuboff (2019), is one of the best-known terms to describe how these different aspects of data come to be connected in a new mode of valuing. Zuboff argues that the ability to predict and to shape human behaviour even in its most mundane and minute aspects is the source of value in surveillance capitalism. Digital technology makes it possible to accumulate data from the personal sphere of individuals, which is a foundational element of the transformation into profit. Surveillance capitalism takes our behaviour as a resource and turns it into a commodity. Zuboff argues that this is a deep and invasive form of exploitation that threatens our integrity as individuals and turns our lives into behavioural data for a few powerful companies to exploit. The critique of surveillance capitalism and how it generates value from our everyday behaviours stresses that it is very difficult, perhaps nearly impossible, to maintain individual sovereignty. Individual sovereignty is the right to determine one's future and the right of sanctuary, to be beyond reach of invasive tracking and monitoring. Zuboff therefore defines the main problem of surveillance capitalism as an issue of fundamental rights of individuals to make decisions about how to live their lives, rather than an issue of monopoly position of BigTech or a privacy issue.

One of the main valuation mechanisms of surveillance capitalism is to turn our behavioural data into predictions, which are in turn sold to advertisers and other companies. These predictions are '… prediction products designed to forecast what we will feel, think and do: now, soon and later' (Zuboff, 2019, p.96). The better these predictions, the more valuable they are. Because the accuracy of prediction relies on obtaining ever more data, in real time, this dynamic further fuels data extraction, so that even more of human experience becomes transformed into behavioural data, argues Zuboff. In turn, when our behaviours are mapped out, this makes them amenable to nudging and influencing, especially in the context of ubiquitous computing. Zuboff is not alone in articulating this critique about the link between data, prediction and the curtailing of freedoms. The massive investments and organisation of physical, institutional and political structures in relation to behavioural data actually shape which futures are more likely to unfold and constrain how we think about what is even possible for us to undertake (Amoore, 2013; Mackenzie, 2017). BigTech, the critique goes, is able to pursue this specific kind of value on a very large scale, to the extent that many of us risk losing out on opportunities, freedoms and privacy.

8.9 The value of metrics

A second way in which machine learning and algorithms are important for value is in the way they form a loop with types of data to give rise to '**metrics**'. The use of metrics is not limited to digital settings, but with datafication, the use of metrics has become amplified. Metrics are therefore a particular category of data practices. The term biometrics is also often used: it is a subcategory of metrics that engages bodies in data analytics. Metrics can be defined as a special form of data that matters. They are measures used for assessment. Metrics facilitate the analysis, evaluation and efficient management or control of a broad range of human activities and practices tied to these measurements and to how we value them (Beer, 2015; Van der Vlist, 2016). Driver ratings are an example of metrics used by Uber. Drivers are rated by riders, and this data matters a great deal in shaping the activities (and source of income) of drivers. The data about ratings of drivers is fed into a decision-making system that vets drivers (drivers must score a 4.6 to continue to access the Uber platform) and that assigns them rides. In turn, drivers behave in ways that are likely to ensure that they have a high enough score to maintain their access to the platform and to their livelihood.

Metrics are valued for the way they seem detached and objective, for what they reveal about phenomena and for the way they can help identify hidden patterns. Metrics can also be gamed so that people start to act according to the measures that will be used to assess their performance. Metrics connect systems of measurement and systems of value, and thereby define what is valued and what are desirable behaviours (Beer, 2015). Metrics therefore perform a phenomenon rather than 'simply' represent it (de Rijcke et al., 2016; Prey, 2020).

One way to summarise these arguments about prediction and metrics is the phrase 'you can't manage what you don't measure'. The other way around of course also holds – there is a tendency to use the data we have to manage, to make the effort of data work worthwhile. However, we have also seen that much of Big Data is not based on measurement, but on the use of proxies – words used on Twitter stand in for 'sentiment', rate of breaking stands in for safe driving, etc. We also noted above that in building AI tools, accuracy may be sacrificed for the sake of efficiency. If we add to this the drive to predict future behaviour based on past data, we end up with a system that is powerful but has serious limitations. By taking these limitations into account, we can better assign value to data and shape what is valuable.

8.10 Conclusion: Making data valuable

If we speak of data as the new oil, we tend to think of data as simply a resource that is out there. As this chapter shows, to become a valuable resource, all kinds of rearrangements of technologies, practices and institutions are needed. When it comes to the economic value of data, there is a strong association between data and innovation, and platforms are a dominant form that shapes the data economy. In addition, the current value of data is characterised by a context where the actors are global and not easily subject to national regulation, where types of activities are not easily assigned to one sector or another and where the instruments of value creation are concentrated in the hands of a few actors. Because data production and collection are pervasive across contexts and tied to innovation, they are challenging to regulate. Finally, the interactions between machine learning and data are also the locus of value creation, through their production of prediction and of metrics. These two forms are important in how the value of data enters our lives. Algorithms do not operate as neutral tools and they shape a context that orders opportunities, privilege and ideas and that is far from being a level playing field. Finally, the discussion of prediction and metrics brings us to questions of automation and bypassing of human engagement, and of how to relate these increasingly prominent tools to human practices and decision making. This calls for attention to data ethics and for responsible data science, the topics of the next chapters in this book.

ADDITIONAL READING

Birch, K. and Muniesa, F. (eds) (2020). *Assetization: Turning Things into Assets in Technoscientific Capitalism*. Cambridge, MA: MIT Press.

Crawford, K. (2021). *Atlas of AI: Power, Politics, and the Planetary Costs of Artificial Intelligence*. New Haven, CT: Yale University Press.

O'Neill, C. (2016). *Weapons of Math Destruction: How Big Data Increases Inequality and Threatens Democracy*. New York: Crown Archetype.

Zuboff, S. (2019). *The Age of Surveillance Capitalism: The Fight for a Human Future at the New Frontier of Power*. New York: PublicAffairs.

9

DATA JUSTICE AND ETHICS

---- Overview of chapter ----

Summary

In this chapter we introduce data ethics. First, we interrogate what forms of data are more ethically sensitive and emphasise the relationships between data work and various kinds of harms, including infringements of individual as well as collective rights. Second, we focus on the significance of promoting data justice for human (and planetary) flourishing, especially in the face of the considerable social damage being wrought by some forms of data work and the complexity of implementing principles such as 'fairness' in everyday data management. We then examine some ways in which traditional, broad frameworks for ethical reasoning can be used for data work. We show how such frameworks can be brought into the very core of technical data operations. We conclude by emphasising how ethical concerns are part and parcel of all aspects of data work.

9.1 Introduction: From data value to data ethics

Recognising the various forms of value associated with data has an immediate practical implication. It becomes evident that any form of data work involves value judgements, and that such value judgements – whether or not they are explicitly acknowledged – have social implications for which data workers can be held accountable. In this chapter, we focus on **data ethics**, which is the study of what it means for data work to be socially responsible and beneficial to life on earth. As we will illustrate, data ethics is closely linked to the study of **data justice**, which focuses on the situations in which data work can be socially damaging and how those situations can be modified to obtain beneficial outcomes instead.

Data justice is one of the key values upheld within the broader framework of data ethics, and one we pay particular attention to in this chapter since it concerns the concrete situations in which data work happens. In other words, while data ethics encompasses the broad range of reasoning around the harms and benefits linked to data work, and how such reasoning translates into action, data justice concerns the specific circumstances of data work, and how those circumstances affect its social outcomes.

Since it reconfigures aspects of agency and power, Big Data raises particular ethical issues (Zwitter, 2014). Deciding on an ethically sound course of action is not a straightforward task for a data worker. For a start, there is no consensus on what such a course of action consists of; data ethics is grounded on broader ethical frameworks for human conduct, which include different understandings of how life should be lived and therefore different views on what counts as 'just' and 'fair' behaviour. There is also no obvious way to apply broad ethical frameworks to everyday data work; each specific data use needs to be assessed in relation to the specific conditions at hand, an operation that requires a good understanding of the social context in question and the possible implications of data-related decisions. This chapter will guide you through some of the key ideas underpinning data justice and ethics in order to highlight the degree of responsibility that data scientists have for the outcomes of their work.

In previous chapters we have seen how the promises of social and technological innovation in relation to Big Data – which range from driverless cars to energy optimisation and robot doctors – make it easy to overlook or underestimate the problems caused by careless data management. Researchers of what Floridi (2014) calls the **infosphere** – the reality created by reliance on digital technologies for all aspects of social life – are becoming aware of the destructive potential of data and the urgent need to focus data management efforts towards the improvement of the human condition. In Floridi's words:

Information and communication technology yields great opportunity which, however, entails the enormous intellectual responsibility of understanding this technology to use it in the most appropriate way. (Floridi, 2014)

These considerations have enormous implications for data work, given the extent to which such work depends on digitally and computationally mediated interactions. Our objectives here are to stress the severity of injustice and discrimination that can be – and often are – caused by careless and unethical data work, and to point to ways in which you can detect and correct such issues. In so doing, we will argue that the analysis of ethical and social issues is a core technical requirement for any kind of data management and analysis. Understanding this fact is essential when considering the concrete rules, laws and principles used to guide contemporary data work that we review in Chapter 10. Given the focus of this book, we address ethical issues that are more directly linked to social dimensions. Data work also raises important ethical questions with regard to environmental and other-than-human dimensions (Lucivero, 2020; Vinuesa et al., 2020; Coeckelbergh, 2021).

9.2 Which data are ethically sensitive?

Much public discourse on data justice, as well as data protection laws, focuses on the rights of the individual data subject and the efforts required to defend individuals from discrimination resulting from unethical data work. The result is that a lot of attention has been devoted to the risks related to the use of personal data, defined as any information that can help to identify an individual. This includes name and address, physical characteristics such as height, eye colour or blood type, or photographed face. This level of concern is certainly warranted, given the capacity of data practices, infrastructures and tools to amplify existing situations of inequity, discrimination and even abuse against individuals. A striking example is provided by one of the most ubiquitous systems of human interaction with the internet: the search engine. In her book *Algorithms of Oppression*, Noble details several ways in which Google search results end up yielding shockingly sexist and racist outputs (Noble, 2018). For instance, a 2013 Google search for 'women need to' would bring up the following suggestions: 'be put in place', 'know their place', 'be controlled' and 'be disciplined' – a result so dramatic that it was denounced by an information campaign launched by the United Nations (Noble, 2018).

This kind of distortion is due to a combination of highly biased data that are registered by the search engine and the algorithms used by search engines. Many users seem to habitually combine the terms 'control', 'women' and 'need' in their online searches. Those data are then analysed through mathematical models and algorithms that unquestioningly incorporate sexist stereotypes and classifications. What makes these results so dangerous is the ease with which they can be confused with a reliable representation of reality. Google is arguably the most trusted and used search engine in the world; as a consequence of its market dominance and popularity, the results of Google searches are often presented as factual reports. Hence when Google creates an association between 'women need' and 'know their place', these associations are further reinforced because Google orders its findings accordingly, and users can be led to believe that these results reflect reality in some way. This looping effect reinforces discrimination in human interactions with the internet, given the importance of search engines as the main interface across many domains of digital practices. It shapes the type and format of the evidence available from search engines, and therefore the use of such evidence within algorithmic systems responsible for evaluating individual situations. This means that search engines filter information used when assessing whether a woman can be a credible candidate for a leadership job or whether a person of colour is eligible for medical insurance. This raises serious questions around the credibility of online searches and the reliability of automated systems used to check eligibility for services. It also points to systemic problems in the ways online platforms portray and affect social life, which have immediate implications for individuals who are labelled as members of discriminated categories.

Can individuals fight back? As it turns out, the opacity of the algorithms used to power Google searches and other automated systems, as well as the difficulties in tracing precisely which data they are trained on and how, make it hard to identify and challenge the mechanisms underpinning such discrimination. The level of technical know-how required to understand and tackle sources of bias within digital systems is also beyond the reach of most individuals – an issue we will come back to in the next sections. Moreover, complaints against decisions taken on the basis of data-intensive algorithmic systems typically need to be backed up with sophisticated arguments, evidence and formal documents that illustrate the error, as we discussed in the case of Uber drivers fighting back against incidents of racial abuse. This is problematic since discrimination is not always a black-and-white affair, but a subtle and pervasive problem that is not always easily demonstrable. The lack of transparency, technical understanding and accessible complaint procedures constitute serious obstacles to people's ability to question discriminatory practices. The difficulties encountered by individuals who complain about data-intensive decision-making systems become ever more pernicious since algorithms are increasingly used to inform decisions ranging from insurance coverage to tax status, citizenship rights, medical or legal assistance, suitability for a given job and criminal charges. Cathy O'Neill has referred to such oppressive algorithms as 'weapons of math destruction' (WMD), pointing out that:

An algorithm processes a slew of statistics and comes up with a probability that a certain person might be a bad hire, a risky borrower, a terrorist, or a miserable teacher. That probability is distilled into a score, which can turn someone's life upside down. And yet when the person fights back, 'suggestive' countervailing evidence simply won't cut it. The case must be ironclad. The human victims of WMDs, we'll see time and again, are held to a far higher standard of evidence than the algorithms themselves (O'Neill, 2016, p.10).

Discrimination on the basis of personal data is certainly extensive and deeply problematic. Ethical concerns, however, are not limited to data pertaining to individuals. Ethical and social issues also emerge in relation to other types of data used to justify social interventions. Data on the use of transport by the residents of a specific neighbourhood, for example, can be combined with data about local access to green spaces to justify the construction of a park or the approval of a new building, which in turn can have positive or negative effects on specific parts of the population. In this way, data on climate, the environment and the biodiversity of a certain geographical area can have significant consequences for the community that resides in that area. Indeed, environmental and climate data contribute significantly to the production of knowledge pertaining to people's habits and preferences. Some uses of such data pose a risk to collectives – that is, any social groupings brought together by a common location, characteristic or purpose and/or long-standing personal ties. Such data can be regarded as ethically sensitive even in the absence of characteristics that enable the identification of individuals. The awareness of potential damage to collectives, in addition to individuals, is very important as it significantly extends the range of data types defined as sensitive (Taylor, 2017).

Consider the following example. One of us was involved in a project that connected medical data with climate data and data extracted from Twitter. The aim was to understand how the incidence of asthma and other seasonal breathing disorders can be connected to particular climate conditions, by looking at evidence provided by people complaining about their symptoms via geolocated Tweets such as: 'I can't breathe today' or 'would have been a great bike ride without the pollen!'. The use of data extracted from Twitter is particularly productive for this type of research because it is a way to address the lack of data documenting less violent asthma attacks. Patients with mild forms of asthma frequently feel unwell but do not often visit a doctor, so that there is no trace of their condition in medical archives. To check when and where people begin to suffer from asthma, it has therefore proved useful to integrate Twitter data with medical records and data on the characteristics of the relevant landscapes (like weather reports and type of pollen of grass present in the area). Linking such data helps to pinpoint the frequency and features of asthma epidemics as well as their connection to microclimates and local flora. This has significant social implications, since the results can be used by public health authorities to manage hospitals and decide how much resource to devote to asthma throughout the course of the year or to inform spatial planning and management of nature areas.

This project exemplifies the opportunities offered by extensive data integration: the resulting understanding of the seasonality of asthma informs state interventions and hospital investments, thus hopefully improving medical services. Yet, this project also exemplifies the social risks involved in using specific kinds of environmental and social media data, even in a fully anonymised form. The decision to use Twitter, for example, was influenced by the fact that this is one of the few social media platforms that allows (although with restrictions) the reuse of its data for research purposes. Twitter however has a very specific type of usership, which includes largely city dwellers under 50 years old from a relatively wealthy, middle-class background. Data extracted from Twitter therefore tend to poorly represent people living in rural areas with less access to public health services – who are precisely the people most likely to be exposed to seasonal changes in asthma. As a result of this bias, a project that looks like an excellent example of effective and responsible data integration for the public good could end up offering results that discriminate against rural populations and reinforce the existing tendency of many governmental agencies to support large cities and underserve less populated areas. A crucial step towards addressing this problem is acknowledging that even projects that aim to serve the public good and do not seem to concern personal data – or use such data only in a fully anonymised form – can have harmful social effects. In other words, even environmental data can be ethically sensitive, and their potential to cause harm needs to be assessed and monitored throughout their use.

9.3 Data justice: Implementing fairness

The potential for drawing misleading, erroneous and/or discriminating inferences underpins the entirety of digital life. Its effects on society are pervasive and systematic, precisely because of the enormous value assigned to data-intensive systems as sources of evidence for decision making as discussed in Chapter 8. From government to medical services, from the justice system to the production of consumer goods, the information garnered through data science is used as empirical grounding for social interventions. It is often viewed as authoritative, reflecting a simplistic understanding of data as neutral representations of reality (critiqued in Chapter 3). In fact, such reliance on data work highlights the political implications of technical decisions. As Eubanks points out,

> automated eligibility systems and predictive analytics are best understood as political decision-making machines. They do not remove bias, they launder it, performing a high-tech sleight of hand that encourages us to perceive deeply political decisions as natural and inevitable. They reinforce some values: efficiency, cost savings, adherence to the rules. They obscure or displace others: self-determination, dignity, autonomy, mutual obligation, trust, due process, equity. They embody very particular ways of understanding the world, and foreclose more promising visions (Eubanks, 2018, p.224).

It is possible to fight against the tendency to see all data work as infallible and authoritative by addressing instances where data work incorporates bias in uncritical and problematic ways. This is the concern of data justice, which involves efforts to locate and counter the social damage wrought by some uses of data and related analytics. Embracing data justice means questioning the extent to which data work supports and respects life, and committing to use data for the good of human society and the planet as a whole. More specifically, data justice involves paying attention to how data work impacts different social groups, which populations are included or excluded from representation in digital systems and how data science can be used to resist – rather than strengthen – existing forms of injustice and discrimination. The concept of data justice is increasingly used to foster fairness and equality in how people are made visible, represented and treated through Big Data, thus enhancing people's capacity for action (Taylor, 2017). This involves opening the 'black box' of data-intensive algorithms to scrutinise how decisions and assumptions incorporated into digital systems may affect social life (Pasquale, 2015). It also involves promoting **data fairness** by researching how data work can help to treat people in ways that are right and reasonable (Veale and Binns, 2017; D'Ignazio and Klein, 2020).

A general framework to understand justice and fairness within processes of knowledge creation is provided by Fricker (2009), who examines the forms of prejudice built into our ways of understanding the world. Fricker distinguishes between two forms of injustice: first, testimonial injustice, defined as prejudice that makes evidence offered by specific groups less credible than evidence offered by other groups. This would be the case where there is a general prejudice against Black witnesses being reliable; the police may not believe statements made by a Black person. The second form is hermeneutical injustice, defined as prejudice built into language and culture used to make sense of the world, which makes it hard to adequately express and evidence specific forms of social damage. This would be the case where there is no concept of sexual harassment and no legal and cultural recognition of this as a crime; it is then difficult for victims of sexual harassment to bring attention to their plight and for perpetrators to be held responsible.

Within data work, clear examples of testimonial injustice appear in the use of specific types of social media as a preferred data source due to those media being the easiest and least expensive to access. The predominance of Twitter as a source for social media data, due to its relative accessibility compared with other social media platforms, is a case in point. The accessibility of Twitter makes it difficult to question its use as a data source in the first place, and/or to argue for alternative data sources requiring additional investment. This has created a prejudice in favour of Twitter data use within much computational social science. Concretely, this means that methods, software and training tools have been created that target specifically this data source and made it even easier to deploy, especially for newcomers to the field (Sinnenberg et al., 2017). Twitter data thus acquire an increasingly authoritative status, as more and more researchers use them as evidence for their studies or as the basis for innovative products and services. Other data sources become proportionally less credible as the basis for innovation or with researchers needing to spend more time to introduce them to their peers and justify their use.

By contrast, hermeneutical injustice takes the form of privileging specific socio-demographic, cultural or political stances within data sources, as in the politically charged case of migration figures, which are often used as evidence of the success or failure of governmental policies. Following the Brexit vote in the UK, for instance, UK residents with European passports were reclassified as immigrants and their data were therefore included in government immigration counts and quotas, with significant implications for the political narratives built around those data. Another case in point is the above-mentioned case of Twitter data for asthma research: hermeneutical injustice happens whenever a study mentions the fact that Twitter users are based in cities rather than rural areas without however taking those biases into account in the rest of the analysis. Some studies may argue that such biases are not relevant to the study at hand. Attention to hermeneutical injustice however demands that assumptions made around whether or not a given dataset is representative, and of what, be examined empirically and considered in light of their potential implications. What evidence is there that the data collected on urban dwellers represent the views and experiences of people living in rural areas? And what are the consequences of this set-up for the studies at hand?

To help mitigate testimonial and hermeneutical injustice, data fairness needs to be conceptualised in relation to the goals, contexts and stakeholders involved in data work. Considerations of fairness are not 'abstract, constrained optimisation problems, but are institutionally and contextually grounded'

(Veale and Binns, 2017). Testimonial injustice can thus be mitigated by data practices that actively counter existing prejudice about what counts as appropriate/relevant/adequate evidence. Hermeneutical injustice can be mitigated by data practices that leverage diverse sources of knowledge to counter existing prejudice due to ignorance or lack of understanding (Leonelli et al., 2021). This includes applying what Leslie (2019) calls the principle of discriminatory non-harm, which he defines as follows:

> The designers and users of AI systems, which process social or demographic data pertaining to features of human subjects, societal patterns, or cultural formations, should prioritise the mitigation of bias and the exclusion of discriminatory influences on the outputs and implementations of their models. Prioritising discriminatory non-harm implies that the designers and users of AI systems ensure that the decisions and behaviours of their models do not generate discriminatory or inequitable impacts on affected individuals and communities.

What does this entail in everyday data work? Leslie proposes to assess each stage of data work through the following steps:

> Step 1: Identify the fairness and bias mitigation dimensions that apply to the specific stage under consideration. Step 2: Scrutinise how your particular AI project might pose risks or have unintended vulnerabilities in each of these areas. Step 3: Take action to correct any existing problems that have been identified, strengthen areas of weakness that have possible discriminatory consequences, and take proactive bias-prevention measures in areas that have been identified to pose potential future risks (Leslie, 2019).

In practice, this may involve several technical adjustments to data work, such as: modifying the algorithms and models used to analyse data; leaving out or adding data sources to correct for a perceived imbalance; taking time to triangulate results with other models and other types of evidence (including, where appropriate, qualitative studies); exercising care in drawing conclusions from data work; and deciding to discard data whose provenance and representational value may prove dubious.

These adjustments remain difficult to implement. The opacity and highly distributed nature of AI systems make it difficult to allocate responsibility around monitoring and addressing eventual bias and discrimination. Moreover, what constitutes fairness changes depending on type of data work and situation. Many different understandings of fairness may apply throughout data journeys, as the contexts and implications of data use shift and change. This includes the basic understanding and application of principles such as individual privacy, which are valued highly within Western ethics and legislation, but are less relevant in other cultural contexts. Because of this variability and range of applications, it is important to consider the broader question of which ethical visions can be incorporated and reinforced within data work and its outputs.

9.4 Ethics for data work: General frameworks

In this section we briefly examine four general frameworks for ethical reasoning and discuss how to use those frameworks to make ethically sensitive decisions and justify those decisions to others. The frameworks that we will consider are: utilitarianism, deontology, virtue ethics and relational ethics. These are key approaches to the domain of **ethics**, which consists of philosophical reflection on what it means to be a good person. The study of ethics helps to address questions around what should be done and *why* we should behave in a specific way. Ethics thus contributes to decisions around **morality**: the systems of norms and rules that tell us what is right and what is wrong, that is, *how* we should behave. Our focus on four ethical frameworks does not constitute an exhaustive review of ethical debates and arguments: there are many other views and options to consider, which you can research further if you are so inclined (see Additional Reading at the end of this chapter). Rather, we mean to provide you with a sense of what different ethical frameworks look like and how they can be used to articulate and evaluate the social implications of data-related decisions. This is sometimes referred to as the 'macro-ethics' of data work: the ensemble of value judgements and reasoning patterns that underpins the choice of goals, preferences and procedures for everyday data management.

First, we consider *utilitarianism*. This is the view that an action is good if it maximises people's welfare in the sense of making the majority of people who are affected by the action happier than they were before the action took place. When an action heightens the happiness of most people, it is defined as having a high utility. Within this framework, the key questions are: Which people matter when defining utility? What is happiness and how can it be maximised? And does it matter at all if a minority of the people affected are made very unhappy by that action?

The *deontological view*, originating from the work of philosopher Immanuel Kant (and therefore often labelled 'Kantian ethics'), takes a very different insight as its starting point. This is what Kant called the categorical imperative: the idea that we should act so to use humanity, whether in our own person or in others, always as an end and never merely as a means. What this means is that no matter the particular circumstances or the number of people who may benefit from an action, there are some actions that are always right or wrong as a matter of principle. Here you can immediately see the difference from utilitarian ethics. In the deontological framework, if even a small number of people suffer enormously as a result of an action, that action is wrong – no matter how many other people may be happy about it. The use of the death penalty exemplifies the clash between these two views: while in a utilitarian framework it may be justified to condemn a dangerous criminal to death, given that this will arguably benefit the rest of society, in a deontological framework this is unacceptable, since it is categorically wrong to kill someone (no matter the circumstances).

The third ethical framework of *virtue ethics*, inspired by the work of Greek philosopher Aristotle, takes yet another approach. Within virtue ethics, what matters are the values that humans use to define what makes a good person. These can include virtues such as truthfulness, courage, temperance, justice and prudence. This view thus moves away from the deontological focus on duties and rights as well as the utilitarian focus on calculating happiness. Here what one needs to decide is which virtues are best applicable to any one situation, and how they should be applied. Justice in particular is a central virtue for many contemporary approaches to data ethics, which insist on its role as a guiding principle for all kinds of data work (e.g. Vallor, 2018).

The fourth and final ethical framework that we consider here is that of *relational ethics*. In contrast to deontological and virtue ethics, which conceive of ethical reasoning as the job of individuals, relational ethics assumes sociality as the starting point for any ethical evaluation. The emphasis is on the inter-dependence between individuals and their communities, therefore privileging the social dimensions of morality and justice over their perception by individuals. Hence what is good for the individual (like, for instance, the rights to privacy and political representation) is conceptualised as derivative of what is good for the group, such as equity of opportunity and the freedom to nurture affective bonds. Relational ethics emerged from African philosophy and particularly the conceptualisation of humanity as 'ubuntu', which roughly translates into the idea that individuals owe their identity, autonomy and sense of self to their role in society and the broader environment. In other words, relations are constitutive of the very idea of individuality: individuals are what they are thanks to their interconnections with the environment and other creatures (Given, 2008; Birhane and Cummins, 2019). Relational ethics was first adopted in Western research in relation to health work (Gabriel and Casemore, 2009), which is not surprising given the significance of social relations in the provision of medical care.

How to use such ethical frameworks within data work? These frameworks can be very useful in helping you to articulate your reasons for making specific decisions. At the same time, each of these frameworks can itself be interpreted in a variety of ways: you are not being told what to do, but rather the frameworks help you to identify and consider different possible courses of action. Within the constraints in which we live and work, the ultimate choice about which actions to take is up to you, and examining your decisions in light of these different frameworks will help you to explain why you took that decisions and explore which implications it may have.

Consider the following example. A city in my country has suffered a sudden flood due to a storm that caused the riven to overflow. I am a researcher trying to provide government with guidance on what emergency measures to put in place to support the population. I need to decide whether to use data from Twitter and from local phone networks to help me gauge the needs and problems experienced by the local population. I am aware of the fact that the people whose personal data I am using have not given their consent to my research. If I apply the utilitarian view to this situation, I may note that the population of the city will be unhappy if my research fails to address their needs during the emergency. Social media analysis provides effective means to identify those needs, so I may conclude that I should carry on. At the same time, if my research findings turn out not to reflect the needs of the whole population (e.g. the people who are not using social media or whose activities are not appropriately represented by phone networks), the population will also be unhappy. This could be interpreted as an argument for not using these data sources. Now consider the deontological view. One the one hand, as a researcher attempting to produce findings for the public good and thereby save lives, I have arguably a right to use these personal data – and social media users have a duty to let me use their data to help improve the situation in their city. On the other hand, social media users have a right to privacy and confidentiality in their digitally mediated interactions, especially since they are not aware that their data may be used for research purposes, and I have a duty to respect these rights. When moving on to virtue ethics, my reasoning could start by acknowledging that I should be brave and have the courage to use social media data in ways that may help address the emergency. I could then note that I should be prudent too and consider the potential long-term consequences of my data work, such as a possible backlash due to the lack of consent to data use. I could also decide to be truthful and recognise that what people say on social media may not be an accurate reflection of what is actually happening to them, which may be an argument not to use the data. Finally, considerations based on relational ethics can highlight the importance of serving my community, including the inhabitants of the flooded city, and therefore doing everything in my power to produce quick and helpful results, but also the risks posed by my analysis, in case the data sourced turned out not to be as reliable as I hope. Caring for my community may involve me spending additional time double-checking the data and looking for more data sources in order to counter such risks.

This example shows how each ethical framework can be used to justify as well as to condemn the use of Twitter and phone network data for this project. Whenever you engage in data work, you are confronted with many different scenarios of application, and making up your mind about what is best can be very difficult. You need to evaluate pros and cons, and long-term implications of the choices you make. You need to think about who will benefit and who might be hurt by your decisions. Evaluating alternative courses of action through the lens of ethical theories can be very helpful in those respects; you can express the reasons for accepting or rejecting a decision, and learn to take and justify hard choices. What is valuable about thinking with these frameworks is therefore not that they prescribe the outcome of your decision-making process, but that you learn to articulate specific arguments in order to define a course of action. The worst decisions are the ones you are not aware of making. These frameworks may also help you understand which ethical principles are dominant in your workplace, or help better understand how different experts take different standpoints in a debate. It is also important to note that these ethical frameworks are not specific to data work and apply to every aspect of social life. Hence, they help to connect data practices with the beliefs and preferences of data scientists and their communities, and signal the continuity between data work and other types of human activity.

9.5 Ethics in data work: Assessing technical decisions

In this section, we turn again to questions of 'microethics': consideration that can help you apply ethical frameworks and principles within everyday routine data work (Bezuidenhout and Ratti, 2020). To ensure that data are used in the most scientifically and socially forward-thinking way it is necessary to transcend the concept of ethics as something external and alien to research, which focuses solely on its premises and consequences rather than on its content. Ethical assessment should, instead, be performed at every step and become a basic component of the background and activity of those who take care of data, as well as the methods and infrastructures used to view and analyse data (Leonelli, 2016b; Mittelstadt et al., 2016). For instance, protecting the confidentiality and security of personal data is critical both to the study and treatment of research subjects and to the robustness with which such data are linked, integrated and compared (Leonelli and Tempini, 2018).

Hence it is imperative to integrate ethical reflection into the technical decisions regarding the management and analysis of data. First and foremost, this requires that we abandon the idea that in order to make an evaluation on the ethical potential of innovation we need to make a precise forecast as to its potential social impact. It is one thing to spend time and resources imagining and validating how a specific innovation could be incorporated into various contexts – a critical effort for the production of knowledge and technologies that react to social requirements and the socio-economic background of its users. It is quite another to try to accurately forecast, quantify and control precise implications. This is an impossible task given that any technology developed with positive intentions can also be used in ethically questionable ways, depending on the goals and context of the user. An example of this is the extraction of personal data from Twitter, which can be used to improve people's lives but also to monitor their behaviour and develop increasingly predatory and instrumental surveillance and insurance systems (Rappert and Selgelid, 2013). This phenomenon is not limited to data work; all technology is subject to the same concern – often referred to as the problem of 'dual use'. It is impossible to keep track of how a tool is used once it has been produced and distributed and there is never a guarantee that a tool can only be used for good ends. This is also true for research explicitly carried out in the name of the collective good. Consider the medical sector, where the new treatments undergo years of testing and rigorous checks before approval, but there is no foolproof way to guarantee there will be no unexpected side effects, especially over the longer term. The ethical evaluation of data work takes place in a context that is both more dynamic and less controlled than the medical one. Within such a context, the analysis of potential ethical and social implications cannot be expected to depend on the ability to accurately predict the future impact of data work.

For similar reasons, it is also important to cast off the idea that research should only focus on innovation with positive social repercussions. This is not a realistic premise for any type of innovation, precisely because all innovation can be used in a way that is damaging to some part of society and involves an element of risk and uncertainty. Just think about the impact that AI is having on the job market, where millions of people – from taxi drivers to teachers, factory workers and lawyers – see their work become more precarious and their working conditions deteriorate through the deployment of automated systems based on algorithms. While no one would argue for a ban on AI given its beneficial uses in other areas, it is clear that a balance must be struck between avoiding social damage and completely arresting technological development. The acquisition of knowledge always brings both advantages and disadvantages and ethical reflection consists of evaluating how these are balanced out and the impact they may have on different segments of society. In the case of AI's impact on the job market, an ethical approach to data science may suggest evaluating which algorithms to prioritise in the provision of services, which sectors should benefit from such technologies, and which cultural, educational and social resources should be deployed to favour a positive social impact. For instance, this can be done by investigating which skills are needed for humans to effectively interact with the technology, thereby creating new forms of work. In other words, ethical evaluation consists of constantly keeping as close an eye as possible on the circumstances that a certain result could be used in, and on the expectations of users. This knowledge should then be used as a foundation for technical decisions regarding sources, formats, classification and analysis of data. This involves looking out for any potential new users of the knowledge that is being produced, who is excluded and who is included in potential use scenarios, and how the research process can be changed to make the results less discriminating, more sustainable and more or less inclusive depending on the requirement.

Asking these questions does not guarantee that they can be adequately answered. As we saw in the previous section, ethical evaluation does not provide certainty on the future, and can easily lead to compromises and doubts rather than to optimal solutions. Nevertheless, it is a critical step towards acquiring better awareness of the potential social dimension of data management. This greater awareness on the part of those producing, mobilising and analysing data increases the sense of responsibility towards the effects of these processes. In turn, this facilitates public dialogue and increased control over the role of data work in society.

There are various examples of Big Data-led innovation whose negative effects could have been contained had their developers systematically and rigorously queried the impact of the choices made. For instance, Facebook in its early years consumed and resold the personal data of its users without much concern for the potential consequences of this behaviour, becoming a full-blown 'Big Brother' and providing various companies and institutions with citizen surveillance tools. In so doing, Facebook overlooked the fact that this type of social media does not necessarily provide an accurate reflection of the real lives of its users and therefore can produce unreliable and discriminatory knowledge. Facebook also positioned itself as neutral with regard to the contents being shared on its platform, repeatedly and explicitly distancing itself from any of the ethical responsibility usually associated with media outlets and news providers. It therefore disregarded the potential for such a platform to support vicious campaigns of misinformation and fraud, including extravagant claims around reptilian aliens invading the planet, Earth being flat and Western medicine being completely unreliable. The freedom to circulate unverified information while systematically targeting some users for such misinformation has led to forms of social abuse and manipulation, for which Facebook is still struggling to make itself accountable. While this design and positioning initially helped the business grow, it may turn against Facebook's interests in the longer term, given the damage to company image and user trust.

Another notable instance is that of Google Flu Trends, a program launched in 2008 with the aim to use data extracted from Google searches to predict influenza epidemics. The idea was to leverage the fact that many patients search for terms like 'influenza', 'symptoms' and 'fever' long before calling out a doctor. Google was hoping to analyse these data to extract more reliable forecasts than those provided by the analysis of official medical data and declared in 2012 that the program enabled the identification of hotbeds for disease 5 days earlier than national health services. The program, however, failed to account for the various terms that users employed to describe symptoms as well as the quantity of searches that were similar to those of people feeling unwell, but were actually made in a different context. In other words, not enough research had been done on how Google is used and which social contexts would generate search terms related to influenza. As a result, the program ended in failure: in 2013 it was unable to predict a particularly significant epidemic, and in 2015 analysis carried out independently from Google showed that the number of cases diagnosed by Google Flu Trends was twice what was verified officially (Lindstrom, 2016). The knowledge produced was both unreliable and discriminatory towards those who really did have influenza but were excluded from this type of research. Cases like this demonstrate the dangers of producing, selling and analysing data just 'because we can' and not on the basis of robust technical and ethical criteria.

9.6 Responsibilities of data workers

The lack of a clear separation between scientifically and ethically justified decisions is particularly relevant when it comes to data work. Any thoughts on the consequences of using old, partial, unreliable or corrupt data are unrelentingly connected to an ethical evaluation of the decisions made during the selection, management and interpretation of data. This is not always acknowledged by data scientists aiming for a given result, often under severe pressure from their sponsors and employers. Even in the instance of the asthma project discussed above, the scientists involved struggled with a deep tension: on the one hand we wanted to produce accurate forecasts both from a social and scientific standpoint, on the other we were frustrated by the constraints linked to data ethics and the efforts required to analyse in detail the sources of the data that we were authorised to use, what type of bias those could include and how such bias could be made visible to other researchers wishing to replicate the research. All this took time and resources, which strengthened the rigour of our findings, but also inevitably lengthened the project. It also made our results less sensational and generalisable. Indeed, our research group would have been more successful had we proposed a simple, fast and non-ambiguous solution to predicting epidemics via automated analysis of Big Data without worrying about potential exceptions or about the discriminatory power of this instrument against the people who are not being represented. Being honest and careful about the conditions under which projections can be reliable makes for scientifically sound and socially responsible research, but this is not always incentivised by the systems within which data workers operate.

This tension is likely to be familiar and understandable to any data worker given the way that research is financed and treated in both the public and the private sector. Influenced by hyped expectations around the power of large-scale data mining, backers of Big Data research expect speedy results with high economic impact – often without considering the time and effort required to verify the reliability, representativity and social impact of data, as well as developing and maintaining adequate standards and infrastructures for data curation. Most of the current systems of incentives for data work do not reward the careful and resource-intensive procedures required for technical and ethical evaluation. What does this mean in terms of accountability for the damaging social effects of data work? Are individual data scientists able to make any change within such a hostile landscape? Is it fair to expect data scientists to take responsibility, given the constraints imposed on their work and the often stringent demands of their employers and customers? In other words, does the responsibility for unjust uses of data science lie exclusively with those in charge of systems of data governance and related incentives?

As we pointed out in Chapter 7, data work is strongly framed by institutional and cultural conditions, with specific systems of data governance in place to control and limit the agency of data workers. There are also important questions emerging over the actions of non-humans like robots and autonomous algorithms ('infraethics', Floridi, 2015), and whether these artificial systems carry responsibility for their 'decisions'. And there are those who argue that the main arbiters of data ethics data use should be the people from whom data are extracted (the data subjects). For instance, the Global Alliance for Genomics and Health (GA4GH), an international coalition of academia, industry and patient groups, has advocated that it is each individual's right to donate their genetic data to science if they so wish, and thus to contribute to the advancement of biomedical knowledge. This idea is sometimes linked to the adoption of the 'right to science' as a fundamental human right (Shaver, 2010; Vayena and Tasioulas, 2015).

Given all this, responsibility for data work and its outputs is unavoidably distributed. What you decide to do with data you are handling in a work setting will never depend solely on you: you will be accountable to employers, governments, industries, collaborators and data providers, and each of these stakeholders will have their own understanding of what should and should not happen with the data. This does not mean, however, that the conditions of data work are completely determined by the social context and data scientists have no agency (and, accordingly, no responsibility) over the goals and modalities of data work. As we discussed in the case of regulations (Chapter 7), institutional demands do not typically determine every aspect of data work – especially the most detailed technical decisions, where data scientists are uniquely positioned to understand what is being decided, what its potential implications are and what alternative courses of action there may be. For instance, many employers care about data not being shared with potential competitors, but do not care about how data are stored as long as they stay within the company. Yet, whether data are stored with appropriate metadata and in ways that guarantee the anonymity of clients can make a big difference to future data work. Or, in the case of health data donation, we have seen how difficult it can be for data donors to assess what implications data sharing may have for themselves and others in the future, given the complexity and opacity of data journeys (Kaye et al., 2015).

Thus, while responsibility and ethical culpability are distributed and constrained by institutional and governance conditions, individual data scientists retain some accountability over their work and can play a significant role in ethical reasoning and assessment. A first step in this direction is to abandon the myth of neutrality that is attached to a purely technocratic understanding of what data science is as a field – a view that depicts data science as the blind churning of numbers and code, devoid of commitments or values except for the aspiration towards increasingly automated reasoning. Data science is sometimes regarded as a methodological field, a sort of generalist toolkit that can be credibly and reliably put to the service of a vast array of goals. And it is certainly possible for data scientists to behave in this way, by taking no interest in the broader context and political interest underpinning their work, and churning numbers for the highest bidder. By contrast, an important task for data scientists is to identify and regularly reassess their **space of agency** – that is, the space within which they can actually make decisions and take responsibility for the implications of those decisions. Another important task is to critically consider the broader work context, even when there seems to be nothing one can do to change one's goals and conditions of employment. In fact, many employers are interested in analysts who are able to highlight potential problems with their data policies and related practices. The ability to contextualise data work can also help choosing a job in the first place, especially when you ask yourself 'why would I want to work with this organisation?'.

When engaging with such ethical assessments, dialogue with others – and particularly with people with different experiences and expertise than one's own – is a crucial source of learning. The rush to offer solutions is not an excuse to avoid any consideration of the social embedding of those solutions. Data science thus needs to work collaboratively with domain experts and relevant communities to forge socially beneficial solutions. In other words, the mindset of data workers needs to include the social, and to do that beyond mere wishful thinking about 'best scenarios' in highly idealised settings. This means imagining and caring about how the outputs of data work can affect the world. Many data scientists are committed to devoting their efforts to the public good; fewer attempt to explicitly articulate and question their own assumptions about what the public good involves, for whom and under which conditions. Conversations beyond one's own team, discipline and habitual peers can contribute to this questioning and to the ability to imagine alternative courses of action. For any given context, there will be a cluster of experts (which may include social scientists as well as humanists, community representatives, professional figures and civil servants among others) who possess an evidence-informed understanding of that specific context. This knowledge is precious and relevant when assessing how a given technical solution may or may not work within a given socio-cultural setting. And indeed, history demonstrates how creative uses of data can be spurred at least as much by confronting the concrete characteristics of real-world situations as they are by pursuing highly idealised visions of the future of humanity. Consider the frequency with which revolutionary discoveries have emerged from highly applied data analysis that took social problem as its starting point, ranging from the rise of computing from Second World War cryptography to the birth of epidemiology from the mapping of cholera transmission on the street maps of 19th-century London.

A particularly relevant moment for data scientists to reflect on the social dimensions of their work happens before data journeys even start; that is, when evaluating whether to participate in a given project or call for funding. Agreement to contribute to a particular programme of work does imply some degree of acceptance of the overall goals and methods set by the project or funding call in question, including the all-important choice of admissible data types. In turn, the choice of whether or not to contribute to a particular research agenda involves some degree of responsibility in prioritising that agenda over and above other possible topics (Garcia et al., 2020). Another crucial moment is project planning. It is often at this stage that decisions are made around which sources and experiences to engage, and whether and how to identify social impact and relevant communities. These choices are constrained by logistical and institutional conditions, including whether or not certain types of expertise are accessible (e.g. whether representatives of a given patient group are available for consultation) and whether there is scope to engage in such dialogue given the timeframe and resources of the project. Yet the ethos and preferences of investigators also play an important role in decisions around how much and how long to invest in identifying and pursuing such engagements. As machine learning and data-driven policy spread across different levels of government and areas of life, the need for such experts will also increase. It is possible for data scientists to be champions of data justice at the same time as they explore new business models and produce novel scientific insights.

9.7 Conclusion: From analysis to action, from rules to power

In Chapter 7, we emphasised that data governance requires ethical data work action at individual, institutional and cultural levels, through regulation as well as an understanding of what to do in everyday praxis. Chapter 8 illustrated the value-laden nature of data work, where technical decisions – including the choice of what counts as data – are shaped by the economic, social and cultural priorities of different social groups, institutions and funders. In this chapter, we showed how the ethical concerns underpinning data work can be conceptualised and articulated in order to enhance the potential of data science for positive social impact and try to avoid damaging forms of injustice and discrimination.

To enhance the accountability of future uses of data, it is useful to build robust accounts of the judgements enshrined in data systems, and to supplement these with explicit reflections on whom they represent, include and exclude. It also helps to bring questions of value to the very heart of data work, rather than pretending that they are 'external' to the process. Perhaps most importantly, it helps to identify and question which power relations underpin and fuel data work. As highlighted by D'Ignazio and Klein in their excellent book *Data Feminism*, this means considering the extent to which data are products of unequal social relations, making data labour visible (including the emotional and affective labour on which data value is often predicated) and embracing the multiplicity of available systems of data analysis and classification (D'Ignazio and Klein, 2020). It also means asking relevant institutions and governments to provide effective legal and cultural frameworks to promote and reward responsible data work.

> This is the time to effect change. There are new data, new tools, and new technologies that can be combined in new ways to create new evidence. There are enough people of good will with enough determination to get things done. Recent legislation has created an opportunity to rethink the organizational data infrastructure. New legislation could take advantage of this golden moment and truly democratize our data (Lane, 2020, p.10).

Working at the micro-level of data practices is a fundamental part of answering such call. Good data science demands venues and infrastructures to care for data, which in turn call for transnational dialogue and support by public and private institutions alike. Good data science involves relentlessly probing the limits and scope of models and algorithms, which in turn demands exposure to – and critical questioning from – the broadest and most diverse audiences. Last but not least, good data science involves a long-term quest to better understand one's social, political and economic context and impact, and to use that awareness to devise and implement an imaginary of data use. In the next two chapters, we will review various concrete steps taken to implement and monitor data ethics and responsible data work, ranging from legal to cultural interventions.

ADDITIONAL READING

Amoore, L. (2020). *Cloud Ethics: Algorithms and the Attributes of Ourselves and Others*. Durham, NC: Duke University Press.

Floridi, L. (2015). *The Ethics of Information*. Oxford: Oxford University Press.

Noble, S. U. (2018). *Algorithms of Oppression: How Search Engines Reinforce Racism*. New York: NYU Press.

Vallor, S. (2018). *Technology and the Virtues: A Philosophical Guide to a Future Worth Wanting*. Oxford: Oxford University Press.

10

RESPONSIBLE USE OF DATA AS EVIDENCE

Summary

The ethical and political issues reviewed in the preceding chapter become particularly acute in situations where data is used as evidence for socially significant decisions, especially those to do with the governance of social and commercial services, and with policy at large. In this chapter, we review the 'evidence-based' movement and its various manifestations in medicine, policy, social order and industry. The scope of these developments shows how important data has become and what is at stake in developing responsible use of data. We then provide an overview and discussion of the main approaches to ensure the responsible use of data, from technical to more abstract tools and concepts. We show how legal frameworks, codes of conduct, design, organisational interventions and consultation can all contribute to responsible use of data. The chapter concludes by stressing that the challenging but feasible task of responsible data use is best approached in collaboration with experts and engagement with stakeholders.

10.1 Introduction: Data matters

In the first part of this book, we discussed datafication as a process of increased data production and circulation, supported by the alignment of four elements (data, capabilities, care and community). In this chapter, we take the element of 'care' as the starting point. How have we come to care about data in this way? How does data contribute to making us care about certain types of evidence? Are there particular challenges associated with using data in this way? How could we better care for data and ensure responsible use? To address these questions, we will further build on the discussions from Chapter 8, where we discussed how data and algorithms are often used to make predictions and decisions. In building and using algorithmic systems, a number of elements are combined – data production, analytic methods, statistical models, computational aspects and hardware – to come to insights. When we apply such insights widely, we speak of algorithmic regulation (Eyert et al., 2020). This is a process where a specific domain of social activity is managed by decision-making systems. When more and more areas of life are shaped by such systems, it becomes increasingly important to understand why and how we put so much trust in algorithms, and to monitor their social impact. Like any other system, it is essential to have checks and balances: no system is perfect and the social situations to which these systems are applied are not static. This is especially important for algorithmically based decision-making systems, since they tend to be opaque and their use can reinforce particular tendencies and leave most of us little recourse.

Let's first consider a case where, perhaps counterintuitively, more data does not lead to better decisions. In a recent experiment pursued by researchers on the AirBnB platform, requests for a reservation from guests with distinctively African American names were 16% less likely to be accepted in comparison with requests from guests with identical profiles but distinctively white names (Edelman et al., 2017). In other words, providing information about the requesting party during this transaction actually made discrimination possible. If AirBnB allows hosts to screen guests, it must be aware of the effect of providing data about names and profile photos to do this screening. It's important to note that this could easily be different: AirBnB does not share all information it has about its users. For example, it does not share email addresses or phone numbers of hosts and guests. Other platforms enable users to pursue transactions with self-selected usernames (pseudonyms). The argument can be made that it is hosts who are racists, and not the platform; at the same time, the design of the platform and the way the data circulate are what open up the possibility for racist decision making. This experiment demonstrates that platforms have a role in preventing or facilitating discrimination, based on the way the data are made available on the platform. When AirBnB was faced with the results of this experiment, it announced measures that include a pledge for hosts not to discriminate, but did not change the data circulation on the platform. AirBnB still provides data that make it possible to continue racist decision making.

Another example about data that matters concerns the use of algorithmic decision making in benefits systems. In most countries, there are benefits systems that are meant to provide assistance to citizens to ensure their well-being. This assistance is usually administered by local or national government agencies and can be in the areas of healthcare, unemployment benefits or access to other services. This means deciding how limited funds can best be allocated – very concretely, deciding who can receive these benefits, what their extent should be and for which period of time. Such decisions are increasingly based on automated, algorithm-driven assessments. These tools quickly process input from multiple sources. Besides making decisions on eligibility to receive benefits, these automated systems are also used to detect fraud and to determine how to distribute care (Brown et al., 2020). They are therefore increasingly important in distributing support to the most vulnerable and in allocating scarce resources. Like in the case of AirBnB, how these systems are set up and the data they use (or don't use) matter. In this kind of system, decisions about which data are considered relevant get built into how the system works. This means that what is relevant to a case is decided a priori. If specific datasets cannot be entered into this system, they cannot be taken into consideration.

In addition, it may not be possible for beneficiaries or their carers to even find out which kinds of data are being taken into account to make decisions about benefits. In some cases, it is impossible for beneficiaries to really get to the bottom of how their benefits were denied or approved, because states and/or companies will point to the proprietary nature of the algorithmic system. To explain how the decision was made would be to give away how the algorithm works, and therefore potentially enable competitors to benefit from this knowledge. The argument can be made, however, that in the case of public benefits, algorithmic systems have been developed using taxpayer money in order to do an essential task of the government. This can be grounds for beneficiaries to claim access to how the system works (Brown et al., 2020). In this second example we see that transparency about how decisions are made, as well as accountability, are important elements in putting data to work.

Across these cases, we see that data is increasingly viewed as a resource to inform socially significant decisions. The structures that make this possible are technically sophisticated platforms and complex, algorithmically driven systems. Such structures are sponsored, hosted and used by organisations, and built and evaluated by professionals and researchers. In the remainder of this chapter, we first consider why these systems have proliferated and then examine a range of mechanisms that can contribute to a responsible use of these systems.

10.2 What is evidence-based decision making?

In this section, we explore why and how data has come to be at the centre of decision making. Two related concepts are important to understand these developments: 'data-driven' and 'evidence-based'. 'Data-driven' is a phrase used to refer to a process that is guided by data. It is often used in contrast to other 'drivers', such as theory, hypothesis or experience. 'Evidence-based' is used to label a strategy that relies on evidence from research, usually in a quantitative form and quite often using specific metrics (as discussed in Chapter 8). The expression 'evidence-based' is often related to policy making, while 'data-driven' can be used in business contexts, research or public administration. Evidence-based movements can be observed in many areas of public life as different as medicine, policing and education. This trend towards 'evidence-based' has been growing in the Global North and in international organisations like the World Bank, the United Nations (UN) and its agencies (Rieder and Simon, 2016). This policy trend also shapes interactions between donors and countries in the Global South, where providing data as evidence of need and/or proper use of aid has become an end in itself (Cinnamon, 2019).

Both terms refer to the importance of having a solid, objective basis for decisions so that it is 'the facts', rather than opinion or assumptions, that are central to decision making. While there are differences between these concepts, they share an important assumption about the possibility of separating data or evidence from the process of their production. In other words, both concepts identify an element (data, evidence) that stands as independent and indisputable, and lends authority to insights or decisions that are said to be derived from them. Importantly, the two concepts have been converging, since what counts as the best 'evidence' is increasingly taken to be 'data'. The result is that evidence-based policy or decision making are practices that have data at their core. For example, discussions about the Sustainable Development Goals of the UN are strongly linked to the growing trend of evidence-based policy. They also formulate that what counts as evidence is 'data'. UN documents therefore widely call for evidence-based policy and tie this appeal to the powers of a 'data revolution' (United Nations, 2018). Data therefore play a 'revolutionary' role in constituting evidence to guide policy.

Grounding of policy in data is meant to enhance trust in the policy makers and their decisions. If decisions are based purely on evidence, strictly on the data as a purely factual representation of reality, then these decisions have authority because they are not tainted by opinion or politics. As we have seen in our discussions of the creation of data and of the knowledge production cycle, such separation of data from their use is unrealistic: data are always part of a process of production, collection and selection, and all these steps contribute to making them usable and meaningful. This insight is important to understand how data can be useful for policy or decision making. While the phrase 'data for policy' implies that one is in the service of the other, the mutually constitutive aspects of this relationship are crucial. For example, the kinds of categories or scales that are relevant for policy makers are also implemented in data collection and data analysis in order for the data to be appropriate to policy making. In turn, policy interest in the data ensures that there is a mandate and resources to maintain data collection practices (Turnhout et al., 2016; Beaulieu, 2021). The kinds of policies we have shape the kinds of data that are used: it is more a loop connecting the two than a one-way street from data to policy.

The particular configuration of evidence-based policy that currently dominates is the result of several historical developments. The modern form of trust that relies on rigorous institutionalised and formalised expertise is the result of increased quantification (Porter, 1995; Shapin, 1995). If, in the early 20th century, numbers were meant to be interpreted by experts, this dynamic shifted in the course of the 20th century when precision and calculation according to highly formalised procedures came to the forefront. The numbers, not the experts, were meant to be the source of authority. In the past decades, this tendency towards quantitative approaches and mechanical objectivity has been further transformed as digital data became central as a form of evidence (Beaulieu, 2001, 2004). We saw in Chapter 8 how data becomes more valuable as it is integrated into decision-making systems. The use of such systems shapes how we trust decision making and how readily we embrace algorithmically driven analysis pipelines (de Rijcke and Beaulieu, 2014).

To appreciate the significance of such changes, it is useful to consider that decision making based on data is not the only way to develop policy or to settle difficult questions. Evidence-based medicine relies on the analysis of multiple clinical studies that have been pursued in different settings and in different countries. The outcome of such analyses can indicate what is the best treatment according to the findings of these studies. In contrast, medicine can also be practised by making a decision based on a doctor's training, her experience, and her evaluation of a specific patient's medical history and wishes. Yet another possibility is the use of a decision-making system based on algorithms that might use even more sources of input (electronic patient files, past cases from that same hospital, costs to the insurance company) and come to a recommendation for treatment. In each of these scenarios there are different roles for technologies, doctors and patients, different kinds of input are considered relevant and different kinds of knowledge are valued. This increased trust in quantified and formalised data is the context in which the 'data revolution' and Big Data have come to be part of so many practices. Data as the basis for making decisions now seems a self-evident, commonsensical approach, and contemporary fields of knowledge and power are articulated around data (Bigo et al., 2019). This shared belief in how to know is called a social epistemology – a socially supported and accepted way of knowing (Jasanoff, 2003).

What are possible consequences of data-driven or evidence-based decision making? First, if data enables better decisions, then logically having more data would mean achieving better decisions. As we have seen in the case of Uber drivers, the use of data as evidence may spur the extension of data gathering and of monitoring activities. A consequence of such intensification is the increase in surveillance as requisite to be able to work. A second consequence is that issues of data quality and bias might be even more critical; errors or omissions in data can matter a great deal if important decisions are based on such data. Third, evidence-based reasoning often comes at the expense of expertise. This can be problematic as human judgement is very important for cases that are exceptional and that do not easily fit within a rule-based system, such as for instance when diagnosing a rare disease or policing a new form of crime (such as newly developed methods to breach data security; Cartwright and Hardie, 2013). Finally, when only 'data' can be taken into account as evidence, this means that challenging a system or decision can only be done on the basis of producing other or more data. Think of Uber chauffeurs installing cameras in their vehicles to try and document misbehaviour of riders and thereby undo harmful reviews that disgruntled riders might unfairly have given a driver. However, as noted in Chapter 8, having to produce 'counter-data' may not be possible for individuals, since the means for large-scale data production are not easily available and are mainly in the hands of large corporations.

The ways in which such systems can go wrong have been amply documented and various critiques have also articulated the need for more transparency, possibilities for appeal and greater accountability when these types of decision-making systems are used (O'Neill, 2016; Noble, 2018; Eubanks, 2018; D'Ignazio and Klein, 2020). In spite of these discussions, there are indications that organisations and data workers still find it difficult to address issues like fairness of their systems, often only discovering problems when there are customer complaints or negative media coverage of their products (Holstein et al., 2019). In the rest of this chapter, we focus on different approaches that can help build better systems and make better use of data.

Before turning to this review of approaches, we want to briefly conclude this section with the suggestion to consider how a concept like 'evidence-informed' might be a useful way to think about the use of data in decision making. To speak of evidence-informed decisions is a way of recognising that may other factors are involved in decision making than data alone and that other elements besides data might serve as evidence (Hazard et al., 2020). Concrete examples can be found in different areas. Take the case of policing. We have seen several examples where data were used by police forces for directing further surveillance, and in ways that infringe on individual's rights by attempting to predict patterns in their behaviour on the basis of potentially biased or incomplete information. These efforts to develop predictive policing in turn led to increased interventions and vulnerability of certain groups (Brayne, 2020). In the project Cutting Crime Impact (Gstrein et al., 2019; Zwitter et al., 2020) data are used in ways that do not focus on surveillance. The project explored ways of using data in combination with the experience and expertise of citizens and police officers familiar with the areas under scrutiny. As the various police organisations gained experience and reflected on the value of data for policing, they moved from a focus on the authority of the data-based modelling and prediction of crime, to looking at data as an additional source of information to be integrated into policing practices. This makes it possible to better contextualise data and use human judgement as a complement to data-driven decision making. In such a case, the evidence base considered by policing – and the objectives envisaged by the project – are broad enough to counteract the risk of discriminatory monitoring. Other examples can be found in various approaches to evaluation, where digital systems are used in ways that require interaction between users and systems to check, adjust and complement the available data to ensure their use is meaningful to those whose work is being evaluated (Beaulieu, 2021). These systems also tend to produce maps, networks or relations rather than numerical indicators. Such an approach requires the deep skill of discernment, not only techniques of data crunching. We return to these larger issues in the final chapter of this book. For now, we consider a range of ways of ensuring the responsible use of data.

10.3 Ensuring responsible use of data

If data is part and parcel of many types of decision making across sectors, how can its responsible use be ensured? There is a strong tendency to point to specific aspects of data use and to target them for improvement. Many public discussions of algorithmic bias have pointed to the use of biased datasets as the cause of bad decision making by systems. 'Garbage in, garbage out' is a frequent way of describing this problem. This framing is found in both popular media coverage as well as among data professionals. For example, Holstein and colleagues note that this framing of 'bias in, bias out' also shapes how data science teams try to improve fairness of their products. They focus on their training datasets, rather than their machine learning models 'as the most important place to intervene to improve fairness in their products' (Holstein et al., 2019, p.600). This is a problematic formulation, because it implies that whatever is processing the 'garbage' is a neutral apparatus and not implicated in the decision making. This is not the case. It is helpful to think back to our model of knowledge production and its many aspects (see Figure 4.2). Knowledge production is also always embedded in a specific context. Any of the components and any of the interactions between them can be a source of bias. This is the reason why the remainder of this chapter will consider a wide diversity of measures that can contribute to more responsible use of data in decision making. There are many ways to improve data governance and use and many actors who should be involved to improve the use of data and prevent harmful or unfair use. Our aim is to give a sense of this diversity. Some approaches will be more feasible or more appropriate in different contexts, depending on the kind of work, the type of organisation or the legal framework that applies. Rather than provide a comprehensive overview – something that could easily fill several books – we put forth measures directed at different aspects of data work to give a sense of the variety of types of interventions and safeguards that can be put in place.

The various instruments and mechanisms we will consider have positive and negative orientations. Positive in the sense that they shape what is acceptable in day-to-day practice, that they help data workers deal with difficult cases or provide guidance about which values to uphold when making design and other decisions. A negative orientation would include the instruments and mechanisms that punish actors who have been negligent, have crossed boundaries or lapsed in upholding proper standards.

Before we delve into the specific descriptions, it is also important to note that the use of specific measures is also related to the kind of accountability that is demanded of particular actors. This can change over time – pioneering efforts are sometimes held to different standards of accountability – or vary across sectors – decisions that affect employment or healthcare may be subject to stricter overview than decisions in the entertainment industry or sports. In addition, procedures and instruments often come to the fore in response to scandals or crises, in a context where more accountability is demanded by 'a public'. Several recent crises have sparked discussion about how social media shape information flows and target particular groups (among these, Cambridge Analytica, the storming of the US Capitol building and the blocking of news-feed by Facebook in Australia are prominent). Such discussions are often framed as the regulation of social media companies standing in opposition to free speech. Another important dimension is the extent to which the public sphere is shaped by corporate platforms in the first place.

At present, we see a trend towards stricter regimes of data protection by governments as well as increased anti-discrimination policies being imple-mented by social media companies themselves. In this sense, the expecta-tions of users about systems also shape what kinds of checks and balances get implemented. Taking this interaction seriously makes the work of devel-oping responsible data use even more complex. Since evaluating whether a system is fair can best be done in relation to users' expectations and beliefs. (Holstein et al., 2019), this points to the need for integration, iterative design and co-creation.

10.4 Legal frameworks and formal regulation

Legal frameworks are an important part of data governance, since they are explicit sets of rules that can be enforced. When it comes to data, the main legal frameworks used are in the areas of privacy, intellectual property and anti-discrimination. Other key issues such as the regulation of digital services are only starting to attract systematic legal attention as we write; for instance, through the European Digital Services Act released in December 2020. This may seem strange given the importance of having legal tools to monitor and constrain the activities of digital platforms. Yet data work and related tech-nologies are allowed to move at a much faster pace than the law, which makes it difficult for legal frameworks to encompass all relevant social implications of these tools. It is not self-evident that technology should 'move fast': com-pare this dynamic of accelerated development and release to the careful test-ing and regulation of drugs, requiring extensive testing to determine effectiveness and safety. Moreover, as we have seen in Chapter 8, many data-intensive actors operate across sectors; for example, Uber straddles informa-tion/communication and transport. Such actors are less obviously subject to regulation because it is less clear where their activities fall. In Chapter 7, we discussed the extent to which property issues around data can shape its circu-lation, while in Chapter 8, we discussed how the ownership of (personal) data has become much more complex due to datafication and extractive practices of BigTech. In terms of responsible use, the result is that existing frameworks have limited impact with regard to uses of data in decision making. Many legal scholars, activists, regulatory bodies and non-governmental organisations (NGOs) are actively seeking to generate new legal frameworks and improve existing ones to make legal recourse more effective.

One such recent effort has been the development and implementation of **General Data Protection Regulation (GDPR)**. It is a prominent European-wide regulation that came into effect in 2018 and takes data protection as its focus. This is a much broader focus than privacy alone because it addresses not only privacy breaches but also aims to help prevent unfair consequences of data management and use. This regulation includes obligations for parties that collect and process data and forces them to proceed carefully with regard to ensuring informed consent, the need to document the impact of their systems and to provide access for subjects to their data. GDPR also makes the reporting of data breaches mandatory. Under GDPR, personal data must be purposefully collected for a legitimate purpose, limited to that purpose and cannot be processed for other purposes that would be incompatible with that original purpose. Data must also be processed in a transparent and accurate way, and it must be possible to rectify data. These many requirements address different parts of data work.

It is important to note how personal data are defined under GDPR: data is any information relating to an identified or identifiable natural person. Another important concept is that of the 'data subject'. This is the person whose data are being collected, who can be identified directly or indirectly by name, location data, online identifier or facets of their identity such as physical, physiological, genetic, mental, economic, cultural or social identity. This broad view of what it takes to identify an individual has implications for many kinds of 'data' – voice recordings, fingerprints, photographs, IP address – which may fall under the scope of the legislation. In addition, GDPR does not only focus on kinds of data but also on what you can and cannot do with them. An IP address might not reveal much on its own, but when linked with data about financial transactions and location data, such a combination might lead to making one identifiable. This is the reason why GDPR also stipulates the kinds of processing that are allowed. According to GDPR, data subjects must explicitly express their agreement with the processing of their personal data. The data subject must consent in a way that is freely given, specific, informed and unambiguous. The organisation that collects and processes data is responsible for this consent and must prove that it was properly obtained. It is also responsible for enforcing the so-called 'right to be forgotten'. This means that a data subject can ask for their data to be deleted.

Social interaction on data production, management and interpretation systems are a critical foundation required to establish what is ethical for whom and in which contexts. GDPR requires anyone reusing data to accurately document the way it is managed and set up and maintain a two-way conversation between those analysing the data and its objects. In preparation for GDPR many data managers in both the public and private sector have been forced to find ways of improving communications with their users. In some ways, it may be easier for larger organisations (than, for example, small- or medium-sized enterprises) to tackle this. This is a hefty exercise that slows and limits the production of knowledge in the short term, but that has huge potential to improve its quality and social impact in the long term. It also proves how implementing social dialogue on data management is easier to enact at the database construction phase than retroactively, once it has already been built.

From this short description, we can already see that this framework is comprehensive (applies across many aspects of data work) and that it articulates the responsibilities of the data processing bodies very explicitly. GDPR is relatively new and some elements are still being worked out. For example, there are divergent legal opinions as to who is considered a data subject whose data is protected. In one interpretation, the data subject is anyone who has residency in the European Union (EU) or is a citizen of the EU whose data are being processed, regardless of where the person is physically located. This would mean that the data of a German living in California and making a purchase from a US-based company would be protected by GDPR. So far, GDPR has mainly been applied to situations where EU-based organisations were processing data from EU citizens. Like any legal framework, the writing and adoption of a set of laws is not the end but the beginning, and much depends on how the laws are applied and enforced.

In order to be effective, organisations that process data must comply with GDPR and it has to be enforced. GDPR states that fines can be imposed on organisations that do not comply – up to 4% of the annual revenue of companies if they violate regulations for data collection, processing and use. With regard to compliance and enforcement of GDPR, there are only the earliest indications of its functioning since its implementation in 2018, but the number and levels of fines have not been radically higher than under previous legislation (Wolff and Atallah, 2020). Finally, GDPR also includes the creation of a new role within organisations, the position of data protection officer, which may help increase awareness and expertise about data protection.

10.5 Codes of conduct

Codes of conduct or codes of ethics articulate the obligations of members of a professional group. The main actor, in the context of a code of conduct, is the autonomous professional who must act ethically. One of the best-known is probably the Hippocratic Oath taken by doctors, whereby medical practitioners promise, among other things, to do their best for the wellbeing of their patients, seek counsel with other doctors where needed, take responsibility for life-or-death decisions and respect the privacy of patients. Many of the experts working with Big Data are keen to highlight the contradiction between the complexity of managing data and the expectation that it should provide reliable knowledge easily, swiftly and in a socially acceptable way. For this reason there are engineers, IT specialists and archivists who insist on the adoption of a code of conduct for data science that encourages those analysing Big Data to take responsibility for the potential social consequences of their choices (Sample, 2019). This would include the obligation to engage in conversation with other stakeholders in order to identify as efficiently as possible what these consequences may be. There are various interesting examples of what this code may consist in are provided by Boyd and Crawford (2012), Zook et al. (2017) and the feminist manifest-no (https://www.manifestno.com). A practical example of such self-imposed behaviour is the moratorium launched by the Ada Lovelace Institute in 2019 concerning the development of automated facial recognition technologies, which was supported by several companies (including Amazon, Google, IBM and Microsoft) as a key move to safeguard the human rights of data subjects (Roussi, 2020).

There are severe constraints to the effectiveness of a code of conduct. Unlike medical professionals, data scientists do not have a strong professional identity, due among other reasons to the variety of training paths and disciplinary backgrounds, as well as their diverse roles and placements within social sectors (see Chapter 5). This may limit the effectiveness of such a code and the possibility of holding each other accountable to it. However, codes of conduct do provide a point of reference for individuals and help spark or sustain discussions within organisations about ethics. They can also serve as a useful benchmark for ethical behaviour in case of disputes (Stark and Hoffmann, 2019). Finally, codes of conduct can sometimes be a first step towards a legislative framework: the moratorium on automated facial recognition technologies, for instance, was highlighted and implemented as an outright ban by the European Commission in 2020, and is under consideration to become a part of UK legislation at the time of writing (2021).

10.6 Computational metrics and design

Recently, the growing attention to moral values like fairness, often noted as a guiding value in codes of conduct, has led to a technical implementation. Data scientists and computer scientists have begun to ask how fairness can be translated into metrics that would enable them to evaluate algorithms and spot problems or opportunities for improvement (Stark and Hoffmann, 2019; Mitchell et al., 2021). Such measures look a lot like metrics for other aspects of systems – efficiency, for example – that we are more used to measuring. This means incorporating a definition of fairness in tests of systems, in order to come up with a quantitative assessment of whether a system is fair or not. While well intended, such an approach may focus too narrowly on testing certain aspects of the architecture of systems and miss many issues that arise when the system is in actual use.

Another set of practices that connect responsible decision making and data focus on the design phase of algorithms and systems. One popular approach is the use of 'red teams'. In this approach to the review of designs or testing of products, a group is designated to take on an adversarial role. The idea is that such a 'red team' can help point out unintended consequences or co-optation of products for negative purposes that might not be visible to the team who has invested in the design of a system of product (Moss and Metcalf, 2020). A related strategy is to pay attention to design as a site where irresponsible use of data can be avoided. The attention to design is often linked to ethical stances and the connection made is one of causation: poor ethics lead to bad designs, which lead to harmful products (Greene et al., 2019). Concern for proper design has been pursued in different, sometimes opposing directions. Some have attempted to create algorithms that are blind to some differences between groups – for example, race. There have been very sophisticated attempts to develop algorithms that do not take race into account, and that also avoid features that might function as proxies for race – postal code or family composition, for example. These attempts aim to ensure that algorithms will not reproduce differences that form the basis for discrimination. A contrasting approach to erasing differences has been to try to improve algorithms in ways that make them more effective at handling difference. An example of this would be to seek to develop facial recognition algorithms that are able to recognise both dark-skinned or light-skinned faces, in contrast to the tools developed so far that are effective at recognising light-skinned faces, thereby reinforcing the norm of whiteness (Buolamwini and Gebru, 2018).

Another approach has been to try and implement data analysis methods that are more targeted, thereby narrowing the opportunities for excessive use of personal data. The growing focus on ethical matters is proving a stimulus to the development of new types of algorithms and data analysis technologies whose objective is to enable the identification of data that are potentially useful for research, without mobilising large volumes of data that are regarded as sensitive or personal. One such example is the development of algorithms and agreements between databases that enable the connection and analysis of certain features of data stored in different locations without accessing the entire database. This enables the analysis of the data without sharing and moving those data from one place to another, thus retaining control over who accesses the data and the objectives towards which they can be used (such as the Data-Shield example by Burton et al. (2015)). A focus on design as site of ethical work is important, but the larger context in which design work is done also shapes it. Business decisions are also the site of ethical work, and responsible data work is also shaped by the choice to deploy surveillance, extraction or 'attention hacking' as we explored in Chapter 8.

10.7 Organisational and cultural interventions

AI and Big Data analysis have developed in the private sector, in an area that has been labelled BigTech. The actors in BigTech have been characterised as special kinds of organisations, having business models that also shape their use of data – infamously, 'move fast and break things' – and operating as monopolies (Tenner, 2018). What an organisation values and how its culture fosters and rewards particular ways of working are also important in shaping responsible data work. At this level, there have been a series of calls to foster better data cultures, and these have been answered by the crafting of 'value statements'. These contrast with codes of conduct since they are directed at organisations rather than professionals. Examples of such value statements include The Partnership on AI to Benefit People and Society, The Montreal Declaration for a Responsible Development of Artificial Intelligence and OpenAI. Across these instances, the roles of data science and data work in shaping society are emphasised, along with the articulation of the responsibilities to society that come with such roles. Like codes of conduct, such declarations are not enforceable. Nevertheless, their symbolic function can contribute to creating a culture of responsibility within organisations, and spur debate and discussion within and between organisations.

Another important example of cultural intervention, this time aimed both at individual researchers and at organisations, is the development and implementation of principles for data management that are specifically concerned with ethical issues. Perhaps the best-known example is the CARE principles developed by the Global Indigenous Data Alliance, which are meant to directly complement the FAIR principles (discussed in Chapter 7) by offering guidance on how to handle concerns with justice and fairness in data work. The CARE principles include: Collective Benefit, which is the importance of evaluating the impact of a given data intervention on groups and communities and ensuring that this impact is positive; Authority to Control, which points to the necessity to distribute power and control over the data across the stakeholders involved, rather than placing all control in the hands of one party (especially if this party consists of digital platforms or specific data users);

Responsibility, which involves the need to clearly acknowledge who is being held responsible when data work goes wrong; and Ethics, which is a broad invitation to monitor the social and moral implication of any kind of data work. Another, similar set of principles was proposed by the Research Data Alliance in order to offer guidance towards the development and maintenance of data infrastructures. These are the so-called TRUST principles, which stand for: Transparency, which is the need to make data operations as easy as possible to understand and scrutinise; Responsibility, in the same interpretation as the CARE principles; User Focus, which involves prioritising the needs, skills and concerns of users over the wishes of infrastructure developers; Sustainability, which implies attention to the long-term prospects and environmental impact of the infrastructure; and Technology, which is the importance of keeping an infrastructure up to date with evolving software and hardware requirements.

Informed by such principles, there is an emerging trend towards public and collective use of knowledge and infrastructures. A number of centres and initiatives have these values at their core. Among them are the Centre for Technomoral Futures at the University of Edinburgh, the research line on Digital Infrastructures for the Public Interest, Stanford Center on Philanthropy and Civil Society (PACS), the PublicSpace coalition, Amsterdam and the Institute for Digital Public Infrastructure at U. Mass Amherst in the United States. These strongly intersect with the 'open' movement (discussed in Chapter 7) and with issues of governance and accountability in the public sphere. This trend is also visible in various programmes that prioritise a responsible approach to research and innovation or to human-centric and trustworthy data technology (European Commission). To instantiate these values, infrastructures, tools and processes around data need to be imagined and built differently, and these initiatives seek to create the conditions for this work to get done.

10.8 Institutional Review Boards

At the level of institutions, a more formal mechanism is the Institutional Review Board (IRB). These boards originated in research institutions (especially medical ones), though since the mid-1990s, they have been widely implemented across Anglo-Saxon universities and later spread more globally. These boards typically review research proposals to ensure that no participants will be harmed by taking part in the (experimental) research. The role of these boards has been extended beyond medical experimentation to consider all 'human subject research' and are often considered the main site of ethical work in research settings. The origins of IRBs also explain their focus on possible harm *to individuals* due to the research method applied. They are consulted at the start of a research project and do not consider possible broader effects or consequences for populations (or society) – the kinds of concerns that are associated with algorithmic discrimination, for example. This means that many concerns about algorithmic decision-making systems or about types of data collection are not addressed by IRBs. When IRBs are set up in corporate settings where an iterative mode of work is the norm, a thorough review at the start of a project may not take into account the further steps in the process of development (Moss and Metcalf, 2020). Many small decisions are made along the way, not only at the start of a project, so when should an IRB be consulted? Some organisations are experimenting with using a mode of work based on 'agile methods' (a mode of software development based on iteration and short and frequent consultation between members of a development team). The idea is to apply this mode of work to ethical issues. Using a mode of work that is familiar in the IT sector can help orient a team to new issues. For example, it provides the opportunity to rethink consequences of the system being developed or to broaden one's view of potential users in the course of the development process.

10.9 Social participation and slow science

Another strategy to contribute to more responsible data work consists of the adoption of deliberative processes where technical decisions made by those analysing Big Data are based on social consultation. This type of consultation may appear utopian to specialists and professionals who often work with scarce resources and under significant financial and time pressure. At the same time, the establishment of procedures that include extensive social debate on the treatment of data is an immediate and constructive way of exploring the ethical and social implications of new uses of data and of decision-making systems. Such consultations draw on the help of those very people who would feel these implications directly. In the world of digital services and social media, users are regularly canvassed for their input to help improve the quality and usefulness of applications. The idea is to extend such consultation to more complex ethical issues and to create bottom-up dynamics that feed the creation of new digital platforms and services and guide their implementation (see, for example, the project Our Data Bodies; https://www.odbproject.org). Getting feedback on ethical elements in Big Data analysis, however, presents two obstacles: one relates to the lack of incentives to get people who are already weighed down with other responsibilities involved in the development of digital systems, the other consists of finding intelligent ways to involve people without a data science background in taking decisions that may appear incomprehensible to anyone outside a restricted group of information technology experts.

The relational view of data can help to overcome these obstacles. We have seen in Chapters 5 and 6 how in the world of data work there is no single role that has a perfect and complete understanding of all the systems used to manage data. As a result, we are in a situation where many people with different points of view and skillsets must cooperate to develop a system that works organically and responsibly. Involvement of users requires the development of communication channels for analysts to debate with potential users so as to help them examine and question their work. This debate should be as unfettered as possible in order to encourage an equal exchange between technicians and the public and encourage technicians to change their digital systems in line with the needs and objections that emerge during the debate. There is no symmetry between all participants in this type of exchange, nor any guarantees that communication will work well and result in reciprocal learning. Yet, there are precedents in highly controversial areas like animal testing in the UK, where the government and the scientific community worked to develop debate arenas where researchers could discuss their reasons for using animals in research and receive suggestions to reduce their number or to improve their treatment compatibly with research objectives. In addition to this, all projects that use animals are regularly examined by government inspectors with a background in biology as well as ethical and legal issues. The regular interaction between inspectors and researchers is not just intended to ensure that rules are followed but also to encourage reflection on the way the research is developing and the potential impact on the animals used as well as the elaboration of transparent explanations of the choices and compromises thus made. The result is the development of occasions when researchers can momentarily suspend their fast-paced activity and take the time to ask themselves how their research methods and impact can be improved such as changing the treatment of animals based on the results obtained to date. This example and its importance to data science is analysed in Leonelli (2016b). Another inspiring example from the UK, this time in the area of data work. is the Secure Anonymised Information Linkage (SAIL) database in Wales. SAIL was set up in the mid-2000s to safeguard and anonymise sensitive data used for medical research, but has gradually evolved to become a centre that organises consultations between the various types of experts, mediates between the needs of patients, doctors and researchers, and provides consultancy for scientists on what research to carry out with such data and in what ways. Its role as social mediator had enabled SAIL to morph from a purely instrumental resource into a critical part of the research process (Jones et al., 2014; Tempini and Leonelli, 2018).

Certainly, different stages and situations in the development of Big Data require different efforts by those involved but starting a debate and ample social cooperation are critical to the management and analysis of Big Data. This type of encouragement to take moments for reflection and evaluation of their work may seem banal, but is actually revolutionary in relation to the incentives and institutional or financial structures in which Big Data work is typically carried out. It also contrasts with a popular interpretation of the BigTech ethos discussed earlier ('move fast and break things'). Big Data is far too often seen as a tool for massively accelerating the production of knowledge without scruples. The establishment of procedures that encourage the ethical management of Big Data can instead help improve the reliability of the knowledge produced, its methodological value, drive data workers to take more responsibility, improve the sustainability of databases and encourage a focus on using data towards truly innovative objectives. Rather akin to the principles of the 'slow food' movement, 'slow science' provides a valid alternative to the existing way of using Big Data. Such a slow approach can help remedy the increasing alienation between data work and 'society', and better align the preferences, requirements and challenges of using data as evidence.

10.10 Conclusion: Responsibility, monitoring and trust

From regulation to new coalitions, we have reviewed a varied set of tools and measures to help ensure the responsible use of data as evidence. None of these solutions is ideal or universal in its impact, and each makes more or less sense depending on the type of data and social or cultural context. This diversity is crucial: responsible data use is the outcome of a dynamic process. There is no simple fix to problems that arise from the collection and use of data. To address problems is a practice, and data workers can draw on a number of tools, techniques, institutions and codes available to guide them (and also to constrain their behaviour, whether they like it or not). Producing responsible data science is feasible, and typically requires collaboration and professional support to deal with its many aspects. To close this chapter, we address two key issues signalled by Mittelstadt (2019). First, responsibility must be assigned across a network of actors that influence the system's design, training and configuration – and, we would add, the deployment and implementation of a system. Which parts of the network should be responsible for what? We believe that many of the approaches discussed in this chapter can encourage social participation in data work and that we need a continuing debate on values and social issues. It is clear that there are times during data journeys – such as when new programming and storage techniques are devised – when data work seems pursued in a separate and independent way from social issues. At the same time, data workers have a great responsibility for the application of their work. In turn, other social segments have the duty to take an interest and be open to debate on the premises and choices made to develop tools for the production of knowledge. There have already been instances when people with no formal training or professional roles in the world of research have contributed in a decisive manner to Big Data analysis;

just think of the development of 'health apps' that quantify our physical wellbeing, data on environment and climate, data produced by social services and demographics, and data provided by patients for biomedical research. It is critical to acknowledge that Big Data experts are not only those who are paid to analyse it and understand statistics and computation. Data expertise also includes knowledge around how data is collected, the conditions and methods, and the implications of its use – this expertise plays a critical role in directing technical decisions. To achieve this greater level of interaction, we will have to change the current tendency of companies that only begin to care about ethics once their long-term viability is established, and once reputational risk becomes a concern (Moss and Metcalf, 2020).

A second issue noted by Mittelstadt (2019) concerns the open-endedness of a system's effects. It is not possible to reliably predict the effects of development choices and this undermines the creation of simple, universal solutions for how to be a good AI developer or use Big Data approaches responsibly. This is why we argue that regular revision of systems and the possibility of recourse are also crucial to responsible use of data. As a system is implemented, there must be a way to appeal to its outcomes and have decisions reviewed. Systems must also be revised periodically, after critical evaluation of their functioning. This process can be called reflexive adaptation, the 'capacity to leverage social learning to detect emerging patterns of discrimination and unfair treatment' (Blasimme and Vayena, 2020, p.762). If we recall the widespread use of data as evidence described in the opening of this chapter, we clearly have much to gain by ensuring that these systems based on data are trustworthy. Trust in these systems is essential to continued trust in our institutions, services and in many forms of collective action. Much is therefore at stake in developing a responsible use of data as evidence.

ADDITIONAL READING

Eubanks, V. (2018). *Automating Inequality: How High-Tech Tools Profile, Police, and Punish the Poor*. Illustrated edition. New York: St Martin's Press.

Lane, J. (2020). *Democratising our Data: A Manifesto*. Cambridge, MA: MIT Press.

Leslie, D. (2019). Understanding artificial intelligence ethics and safety: A guide for the responsible design and implementation of AI systems in the public sector. *The Alan Turing Institute, e-Prints*, https://ui.adsabs.harvard.edu/abs/2019arXiv190605684L (accessed 21 February 2021).

Zook, M., Barocas, S., Crawford, K., Keller, E., Gangadharan, S. P., Goodman, A., Hollander, R., Koenig, B. A., Metcalf, J., Narayanan, A. and Nelson, A. (2017). Ten Simple rules for responsible Big Data research. *PLoS Computational Biology*, 13(3): e1005399–e1005399.

Part V

CONCLUSION: DATA AND THE KNOWLEDGE WE NEED

Summary

The final part concisely reflects on data assemblages as important sites of investigation, accountability and action. It formulates the main insights of the book as five key lessons. We highlight the importance of ongoing reviews of data practices and related engagement. Taking account of the complex, dynamic and situated nature of data work is essential to create better, more responsible uses of data and of the knowledge acquired from data analysis.

Learning objectives

This part will help you to:

1. formulate the importance of responsible data practices in discussion with a variety of professionals and stakeholders;
2. recognise opportunities to reflect on and act to promote good data practices;
3. understand the open-endedness of data work and the need to regularly engage in review and evaluation of data systems and practices.

Data Story 7: Emergency Data Science and the Use of COVID-19 Contact Tracing Apps

When the COVID-19 pandemic emerged in early 2020, governments and public health authorities around the world frantically looked for systems to track the spread of infections in real time. A seemingly obvious solution was to use a technology that many individuals already carry with them whenever they leave their homes and that is capable of geolocating individuals: smartphones. Hence many governments invested in developing data management apps running on smartphones to track exposure to COVID-19. These apps would proceed by collecting data about people's geographical movements and contacts with other individuals. Whenever an individual would test positive to COVID-19, these data would make it possible to reconstruct that individual's behaviour and trace the people (such as family members, friends, colleagues or strangers) who may have been infected by the individual up to that point, based on their presence together in closed places (such as restaurants, offices, trains or places of worship). This knowledge could then be used in two equally significant ways. First, the affected individuals and places would be alerted about the possible infection risk and would be asked to self-isolate (in the case of individuals) or close down temporarily (in the case of closed places), which would help to contain transmission. Second, the data would help public health authorities to track the COVID-19 rate of transmission (the speed with which the virus was spreading) and geographical spread. This could inform an overall understanding of the possible increase of infections for any given area and related policy measures around social distancing.

- *Do contact tracing apps infringe on individuals' privacy?* Can you think of ways in which data about an individual's location and social contacts could be used against them? Are privacy concerns significant regardless of the social context, or are they more significant under specific circumstances (e.g. a particular political regime or for an individual who is subject to social discrimination)?

Privacy was recognised by many countries and policy organisations as a key problem for the deployment of contact tracing apps. In countries such as Germany, where there is a strong commitment to protecting individual privacy, the idea of developing contract tracing apps was received critically by media outlets, organisations and lobby groups across the entire political spectrum. One solution to this debate over privacy infringements was offered by a collaboration between Apple and Google, which proposed to store all relevant data on the phones themselves (a 'decentralised' approach). Keeping data local would enable users to receive an alert whenever their phone detected possible exposure to COVID-19, for instance if the users come into close and prolonged proximity with an infected individual, without however revealing the source of that exposure and without disclosing the movements of the individuals concerned to public authorities.

This solution was widely adopted by countries with high transmission rates such as Italy and the UK. It is notable however that the focus on protecting privacy ended up clashing with other individual rights in this case, such as the right to medical support. By decentralising data collection, these apps did not make it possible to understand the origin of any one alert, or to provide support to the individuals affected (since their identity would remain anonymous). The app was also separate from the data systems maintained by public health and medical services, and therefore could not contribute to efforts to map the spread of disease over the national territory. Such data could have greatly enhanced the ability to control transmission and develop targeted policies. As a result of the local, decentralised storage of data, these efforts suffered from data scarcity and significant difficulties in linking existing datasets.

- *Do efforts to protect privacy necessarily clash with efforts to trace transmission and inform public health?* Could such a clash be avoided? Which experts need to collaborate in order to develop solutions in such a case?

Levels of penetration of smartphones in some parts of the world exceed 100%. This can make it seem like 'everybody' has access to this technology and that it is a good basis for reaching the entire population since the technology is already in use and widely adopted. Yet apps function on specific software platforms and smartphones require connection to reliable infrastructures, including electricity networks, stable broadband and power sources for charging. The quality of such infrastructures varies enormously between countries and between rural and urban areas. There are also inequities in who gets to use such technology, with women having significantly less access to smartphones than men in many countries.

- *What are the implications of assuming that contact tracing apps are accessible to all?* Could the app privilege those who have the latest smartphone models, or who can afford high-data-use contracts? Could the app fail to assist those who may be most vulnerable and least visible? What problems does uneven coverage create for public health and for society at large? How can possible inequities be identified and who should be responsible for this?

A key factor underpinning the deployment of contact tracing apps was urgency. The COVID-19 pandemic was a clear emergency situation: a global disaster that wrecked the economy and livelihood of billions of people, caused millions of deaths and long-term disability, compromised international travel, and damaged social and affective

(Continued)

relations within most countries – arguably resulting in a worldwide mental health crisis. Under such circumstances, data workers and public health authorities were under immense pressure to provide a solution to contact tracing as *quickly* as possible.

- *What difference do emergency conditions make?* Does the need to tackle an emergency justify rushed technical solutions? Do ethical and social considerations stand in the way of an effective and rapid disaster response?

Decisions taken during emergencies often have an enormous social impact both in the short and in the long term. In particular, decisions about which technologies to adopt and how to use the resulting data can shape the future in significant ways and create path dependency. For instance, it is relatively easy to continue using an app created to track population movements during a pandemic even after the emergency has passed. App users will have become used to it and may even forget that they have downloaded that technology on their phones. And there are many advantages to continuing this form of data tracking – not least in order to increase readiness for the next pandemic or other disaster.

- *When does an emergency end and what are the implications for data technologies?* On which basis should decisions be taken on whether to support or discontinue technological solutions adopted under emergency conditions? Who is responsible for such decisions? Who should be involved in reviewing the role of such technology in society? Could the development of 'immunity passports' (certifying that the carrier has been vaccinated against COVID-19) prolong and extend the use of COVID-19 apps? If they are used for travel, could such immunity passports increase the global circulation of personal data?

Some contact tracing apps, such as the French StopCovid, did not follow the Apple–Google proposal and instead developed a more centralised solution that would enable the apps to share some limited data with relevant public authorities. This required consultations with public health authorities as well as international experts in both epidemiological data and data ethics. These consultations ensured that privacy would be safeguarded without compromising the government's ability to support citizens at risk. The consultations took a few months and they were facilitated by existing networks of data workers (such as the Research Data Alliance) that immediately established relevant working groups to help support the pandemic response.

- *How does a centralised approach to contract tracing fare in comparison with a decentralised approach?* What are the advantages and disadvantages of these two models? Are there ways of combining their positive features, while avoiding their negative ones?

The existence of long-standing initiatives around pandemic preparation, led by international bodies such as the World Health Organization, proved essential to the coordination and fast deployment of solutions to the COVID-19 crisis. At the same time, the pandemic highlighted the scarcity of data infrastructures and standards that would enable fast and reliable data sharing within and across countries. Access to relevant data was one of the key challenges confronted by epidemiologists, medical staff and public health authorities involved in the pandemic response.

- *What kind of data work is required to improve readiness for the next global disaster?* What kinds of investments should we aim for, both in terms of training and in terms of infrastructures?

Data story based on:

Blasimme, A. and Vayena, E. (2020). What's next for COVID-19 Apps? Governance and Oversight. Science, 370(6518): 760–762. https://doi.org/10.1126/science.abd9006

Kolaczyk, E., Lee, M. M., Liu, J., & Parker, M. S. (2021). We need a (responsible!) data science rapid response network. Harvard Data Science Review. https://doi.org/10.1162/99608f92.2794e78d

Krige, J. and Leonelli, S. (2021). Mobilizing the translational history of knowledge flows: COVID-19 and the politics of knowledge at the borders. History and Technology. https://doi.org/10.1080/07341512.2021.1890524

Leonelli, S. (2021). Data science in times of pan(dem)ic. Harvard Data Science Review, 3: 1. https://doi.org/10.1162/99608f92.fbb1bdd6

Leslie, D. (2020). Tackling COVID-19 through responsible AI innovation: Five steps in the right direction. Harvard Data Science Review, Special Issue 1-COVID-19

Data Story 8: Combining Data to Advance the Sustainable Development Goals

To address global problems such as climate change, loss of biodiversity and poverty, the United Nations (UN) formulated 17 Sustainable Development Goals (SDGs) in 2015 as part of the declaration *Transforming our World: the 2030 Agenda for Sustainable Development*. SDGs identify broad areas that pose enormous challenges to the survival of humanity and of the planet, ranging from health to food security, gender equality, clean water and climate action. There is much hope that the collection of high-quality Big Data can help to

(Continued)

monitor progress towards achieving these goals. Fulfilling SDGs involves knowing how well different countries are progressing towards these goals. Funding agencies and policy makers are keen to use data science to assess 'what works' among the many programmes supported within this framework. Many efforts focus on improving methods and conditions for data collection and reuse, including relevant data infrastructures, governance and metrics (as discussed in Chapters 6 and 10).

In what follows we focus on the metrics and related data practices. The 17 SDGs have been defined according to 169 targets, for which 234 indicators were selected. Progress is measured according to the indicators, which are things that are considered to be measurable. For example, the proportion of births attended by skilled health personnel is an indicator for the target of reducing maternal mortality, which in turn forms part of SDG 3 ('Ensure healthy lives and promote wellbeing for all at all ages').

Data relevant to each of the indicators is consolidated at the national level by mandated official national statistics offices. It is then sent to international 'custodian agencies' like the World Health Organization or World Bank. These agencies in turn report regularly to the UN on the progress made towards meeting the targets associated with the SDGs.

- *What is the relationship between data, indicators, targets and goals at a global scale?* What are the potential problems with aggregating data at national and international levels on such complex targets? What kind of evidence of progress towards the SDGs do indicators provide? What is needed to produce data in this way? What makes data reliable when brought together from so many sources?

Another target for SDG 3 is target 3.4:

> By 2030, reduce by one third premature mortality from non-communicable diseases through prevention and treatment and promote mental health and well-being.

An indicator for mental health and wellbeing is the suicide mortality rate, which is labelled as indicator 3.4.2:

> defined as the number of suicide deaths in a year, divided by the population, and multiplied by 100,000.

This is the only indicator for mental health that is taken up in the SDGs.

- *How do indicators measure complex social phenomena such as wellbeing?*

What kind of evidence does indicator 3.4.2 provide about mental health? Are there other ways of measuring wellbeing that might be useful to take into account? What would be their advantages and disadvantages? How could local variations in data collection and analysis affect the data?

Providing this kind of evidence is complex, but it is considered necessary in order to verify whether development aid and intervention programmes are actually helping. Data from many projects in many locations are aggregated to provide an overview, often presented as a single figure.

- *What constitutes responsible use of indicators?* In order to use indicators as measurement for progress, how does it help to have an understanding of the practices of data production, sharing, aggregation and circulation used to create the indicator? Can a single individual understand such practices in all their complexity?

Policy makers require evidence to act. Within the SDG framework, the collection and analysis of digital data on a global scale is considered a primary source for policy-relevant evidence. Many national and international agencies therefore invest attention and funding into developing forms of datafication that can help inform SDG indicators and targets.

- *Is the attention and investment in collecting and aggregating data justified?* How and why? How does effort spent on data work relate to effort spent on social interventions? Is there a danger that too much money and effort is devoted to collecting and aggregating data, rather than on the social and environmental needs of the planet?

Some actors worry that indicators have become too important and that meaningful assessment, which is closer to the actual activities pursued, should be privileged over measurement. This critique is also heard in fields other than development, such as science and educational policy. A phrase that often recurs is that the monitoring of development should aim 'to improve rather than (only) prove'. The specificity and rigidity of indicators are at times argued to be in tension with the achievement of significant progress.

- *Can data help support goals in other ways than by producing metrics and indicators?* Is it possible to combine data and digital technologies in ways that are more attuned to local meanings? What would a format like a Data Story add that an indicator cannot?

(Continued)

Data story based on:

Beaulieu, A. (2021). Data practices and SDGs: Organising knowledge for sustainable futures. In M. Hojer Bruun, D. Brogaard Kristensen, R. Douglas-Jones, C. Hasse, K. Høyer, B. R. Winthereik and A. Wahlberg (eds), Handbook for the Anthropology of Technology. London: Palgrave Macmillan.

11

TOWARDS GOOD DATA SCIENCE

———————————— Overview of chapter ————————————

11.1 Lesson 1: 'Data' is a relational category (Chapters 1–4)
11.2 Lesson 2: Infrastructures and data stewardship are essential to extract knowledge from Big Data Chapters 5–7)
11.3 Lesson 3: Data workers must use data sources with discernment and be aware of the risks of discrimination and inequality connected to data (Chapters 8–9)
11.4 Lesson 4: Ethics, security and social responsibility are a fundamental part of data work (Chapters 9–10)
11.5 Lesson 5: Responsible data work requires social dialogue, community engagement and contributions to data literacy

Summary

In this final chapter, we identify five main lessons from this book and point to their practical consequences for data work. While the discussions in this book also address institutional and societal dynamics, these lessons are more specifically directed to data workers as actors on the ground. In conclusion, we stress the importance of fostering data engagement for ensuring that data are used responsibly across all stages of data journeys.

11.1 Lesson 1: 'Data' is a relational category (Chapters 1–4)

There is no data without a relationship: data do not speak for themselves. Data are only interpretable in relation to a network of conceptual, material and social relationships that need to be made explicit to justify the results of the analysis. This explains why different social groups often attribute different value to the same data, depending on their goals and background knowledge. This potential for variation creates the opportunity to repurpose data, which makes data particularly valuable while also making it hard to predict future data use scenarios. It also explains why what counts as reliable data for one individual may not qualify as such for another individual: the very decision of what counts as data is context-dependent. This does not necessarily take away the value of data as empirical evidence. Rather, it means that the value of data as evidence can only be assessed in relation to specific contexts of data use. Many datasets can be put to new uses and have multiple uses at once, but whether or not such diverse uses are appropriate and credible needs to be evaluated on a case-by-case basis.

Practical consequences: data are not neutral facts and their use as empirical evidence needs to be carefully justified whenever data are employed in a new context. The importance of contextualising data also means that data management is critical to data reuse. Data management practices, including strategies to store and annotate data, help to document the provenance of data and the various ways in which they have been processed up until the moment when they are considered for analysis. This is particularly useful in cases of complex data journeys. It also helps to assess the reliability of a given dataset for the specific purpose and context at hand. Ignoring relationships – the context that defines what counts as data – is actually what often causes problems. Data reuse works best when the history of data practices is documented explicitly and made easily accessible to prospective data users. Data management practices make it possible for data to be reused effectively, and therefore need to be funded and appreciated accordingly.

11.2 Lesson 2: Infrastructures and data stewardship are essential to extract knowledge from Big Data (Chapters 5–7)

The accumulation and interoperability of data require a conceptual, material and institutional supporting apparatus in the shape of infrastructure, databases, regulation and adequate training. This in turn calls for specific and substantial resources to maintain and regularly update this apparatus in the long term. Good data storage, stewardship and management contribute to trust and maintaining fairness in data work. Those involved in the analysis of data should take an interest in the workings of databases and in the algorithms used to mine data. By collaborating with others who are responsible for different aspects of data, data workers can critically evaluate the impact of tools, methodologies and classification systems on the knowledge extracted. At the same time, research-intensive organisations need to understand the importance of hiring experts in the handling, governance and analysis of data to help data workers negotiate the complex issues of their practice. Ongoing and long-term maintenance of infrastructure is a requirement for the reliability of data.

Practical consequences: strategies for sharing data should be considered from the outset of data work, and their implications regularly reviewed throughout. The data and associated metadata need to be appropriately and securely stored, and data management plans should be developed in tandem with ethical reviews to explicitly detail these procedures. One approach is to design appropriate management plans and systems architecture for the secure storage and retrieval of the datasets. Data users, and especially large institutions from the public as well as the private sectors, need to invest in responsible data use. This can take the shape of developing reliable standards, venues, infrastructures and training. An example of this is provided by cases where research institutions work closely with national governments, consortia and international associations to develop and maintain efficient systems to care for data. This is emerging in calls for digital public infrastructure and for seeing platforms as utilities, and in plans for federated systems such as the European Open Science Cloud. Data workers need to be given new prominence and recognition in universities and industry given their key role in ensuring that data are used correctly. Professionals in every field should have basic training in the use of technologies, infrastructure and methods relating to data, as well as in the collaborative skills that are needed to work with data effectively and responsibly.

11.3 Lesson 3: Data workers must use data sources with discernment and be aware of the risks of discrimination and inequality connected to data (Chapters 8–9)

Data work needs to build on appropriate sources of evidence, taking account of the novel types of discrimination created by data dissemination and reuse.

The extensive *availability* of data does not guarantee its *suitability*, and users must consider the strengths and limitations of the data in relation to each project. One particular aspect of assessing the suitability of data relates to the context of production. Social media posts, for instance, are neither single, standalone comments, nor necessarily pieces of 'truthful' information. Their production and generation, and indeed any responses they provoke, are part of a complex social process. Common approaches to gathering and processing of data tend to strip them from their context, potentially limiting the interpretation and understanding of the meaning of the information. Clarity in reporting data choices and sampling procedures is one element of the process that enables ethical evaluations of data work. Every instance of data sharing also needs to be evaluated by weighing its value against the potential harm that it may yield, particularly when integrated with other data. Other challenges are posed by the opacity in the storage, display and data provision/distribution protocols. This is particularly relevant to commercial companies though the accountability of publicly funded research is also under pressure. This opacity has serious implications for the reliability and validity of any data work undertaken using those data and may introduce hidden biases. For example, when geolocating data extracted from applications such as Instagram, it is unclear how location is set and measured. This can generate inequalities in sampling and representation, depending on the scale and resolution of location categories used. More generally, as long as we have little information on how algorithms work for specific social media or decision-making systems, it is difficult to evaluate assumptions being made and how they affect the outcomes of data work. These infrastructures and ways of working shape the kinds of innovation and value creation associated with data. The way we work with data determines who benefits from data produced by users, and from its further circulation. These arrangements affect labour, property and profit and can be a source of increased inequality.

Practical consequences: the selection of the data to load into a database needs to be documented explicitly and to be based on appropriate sampling methods. Data workers also need to diversify their data sources as much as possible given the constraints of their projects, and regularly review the appropriateness and relevance of the sources already in use. Efforts to facilitate the mining, integration and reuse of various types of data need to be supported by individuals with widely different expertise, so as to maximise the evidential value that can be extracted from the data. Data practices should be made accountable to those who, while they are not directly involved in them, are affected by them. This means fostering sustainable instruments of data governance, including pursuing and regularly reviewing strategies to document the work being carried out and facilitate scrutiny from other data workers. It also means expanding the range of who is considered as relevant stakeholders and publics. This does not necessarily require making all data of relevance freely available to anybody who wishes to see them. While open access to data can stimulate creative research and novel interpretations, it can also increase the risk of misuse and harm to human subjects. For any given data project, tools and opportunities for review and interaction should be developed to tailor data access, prospective reuse, and evaluation to research needs and goals.

11.4 Lesson 4: Ethics, security and social responsibility are a fundamental part of data work (Chapters 9–10)

We have argued that data workers need to engage in a critical process of reflection on the context of their work, the implications of the positions being taken, as well as the basic questions of whether available data could or, more importantly, should be used. The intentions and assumptions of any instance of data work should be explicitly considered in the planning of the research and adequately reported in the final stages. Though admittedly difficult, the culture, assumptions and commitments already in place in any given data science environment may need to be challenged to ensure practices are lawful and socially fair. There is a wealth of well-established and emerging mechanisms to ensure responsible use of data. These can be methodological (research plans), institutional (Institutional Research Boards) or organisational (value statements), legal (General Data Protection Regulation) technical (fair algorithms) or professional (codes of conduct). These help data workers and organisations take into account the potential consequences of using specific types of data. Awareness of what data may or may not represent should drive choices made at all stages of data work. This is clearly of importance for the primary collection, processing and analysis of datasets, but also applies to the secondary use of data and to the application of any tools developed. It is not possible to separate the ethical value of data from its scientific value. Anyone who is responsible for organising and analysing data also bears some responsibility for the way rules are applied in each specific instance. Fundamental choices made at the point of creation of knowledge from data have a critical impact on the reliability and social impact of the results. Although the different data work is distributed among people with different highly specialised skills, they each bear a part of the responsibility towards the social consequences of the tools and knowledge thus obtained.

215

Practical consequences: every element involved in the production, management and reuse of data needs to be considered in relation to both its technical value and its social and ethical implications. Building safeguards for social and ethical concerns with data reuse can help to make the resulting insights methodologically sound, accountable to and engaged with diverse stakeholders. It also helps to make data work robust to continuously changing requirements, contexts and challenges. This calls for careful planning around the scheduling and impact of data work, taking interdependencies and potential delays into account, and assessing the potential ethical and legal implications of data reuse at regular intervals throughout data projects.

11.5 Lesson 5: Responsible data work requires social dialogue, community engagement and contributions to data literacy

Working with stakeholder and reference groups and, where appropriate, specific publics, can help refine research plans and, crucially, surface unknown unknowns. Public engagement and trust are thus central to data work, and particularly to ensure that knowledge produced through data work is suited to the relevant social contexts. The emergence of the citizen science movement and 'quantified self' technologies calls for increasingly engaged relationships between researchers, data curators and analysts, and different publics. Part of data engagement for which we plead in this book can be labelled 'data literacy'. This includes awareness of the meaning of data and of how conclusions are drawn from it, and the ability to understand data presented in charts or infographics. Even more significantly, the engagement we plead for also includes an understanding of the importance of data as a social, cultural, material and economic phenomenon. This means that concepts like datafication, data journeys, knowledge infrastructure, the data economy and data ethics need to become much more prominent as part of our general knowledge. This is very important in the research world, where different disciplines have different levels of understanding of the implications of sharing and integrating large datasets. There is also a growing concern in society as a whole with the extent to which data circulates and shapes everyday life. These complex processes need to be discussed more widely with more stakeholders and in more precise terms. As we noted in Chapter 8, pitting privacy against innovation does not do justice to both the potential benefits and threats of intensified data use. As a society, we need to foster a higher level of understanding concerning the nature and implications of Big Data sharing and use. This is becoming indispensable to be able to engage cogently with public and private services, technologies and social media, and to demand that they respect human dignity. Public engagement should include the discussion of research plans and modalities of data sharing, raise awareness of how data are being used in data work, and develop new research directions and hypotheses. It is also important for researchers to clarify what is meant by 'public' and 'public good' whenever planning such collaborations. Scholarly work on the information commons, non-rivalrous uses of knowledge and alternative forms of consent is growing, and could provide an effective reference point for setting up engagement practices. Such reflection provides a needed counterbalance to the current tendency towards data appropriation and privatisation. In a world where data is a financially valuable commodity, it is all too easy to transform acts of data sharing and open data regimes into mechanisms for the limitation of creativity, infringement of privacy and theft of intellectual property. The variety of national legislations surrounding the use of personal data and the transfer of materials (such as samples), and the legitimate worries raised by bioprospecting practices, make this landscape ever more challenging and deserving of careful consideration.

Practical consequences: community consultation and social debate need to be incorporated into the procedures for building and maintaining databases and analysis techniques for data. It is beneficial to support institutions, venues and forms of expertise dedicated to mediating between different data users and publics. Data workers should participate in activities aimed at communicating data science concepts and skills and an understanding of individual and collective rights relative to Big Data access, use and interpretation. The creation of arenas for debate and social participation in data-intensive processes needs to be considered and promoted as part of government, university and corporate research. Data workers should be encouraged and enabled to participate in these efforts. The national school curriculum should include practical education and discussion around the creation, management and interpretation of data, while governments and societal organisations need to organise training courses to update different publics and provide opportunities for debate.

GLOSSARY

AI/artifical intelligence is an umbrella term used to encompass a broad range of automated forms of data analysis, typically combining statistics, modelling, programming and computing.

Algorithm in the context of machine learning refers to the operations and calculations performed on data.

APIs (application programming interfaces) are an interface built by platforms that makes it possible to connect applications and share data.

Big Data is a loose term often used to refer to the large and diverse body of data generated by digital technologies through the datafication of human activities.

Big Data empiricism is the belief that data are the best form of evidence to establish truth, to form opinions about the world and to make judgements.

Big Data mythology is a set of inflated expectations around how Big Data enables new, cheap and efficient ways to plan, conduct, institutionalise, disseminate and assess research.

Conventions are standards agreed upon by the relevant stakeholders, which make it possible to retrieve and link data across different platforms.

Correlation is a statistical relationship between two data values. When two values are correlated, we know that when one changes, the other will change as well.

Curation in the context of data work is the process of organising, formatting and annotating data so that the data can be retrieved, shared and reused.

Datafication is the process of turning of objects and activities into data.

Data ethics is the study of what it means for data work to be socially responsible and beneficial to life on earth.

Data fairness involves considering how data work can help to treat people in ways that are right and reasonable.

Data governance refers to the ensemble of regulations, norms and socio-technical systems that enables and directs data work and particularly how, where and why data can travel.

Data journeys designate the movement of data from their site of production to many other sites where they are processed, mobilised and repurposed. Sites can encompass diverse times, disciplines or viewpoints.

Data justice concerns the specific circumstances of data work, and how those circumstances may affect whether such work is socially damaging or socially beneficial (and to whom).

Data mobility is a label for the extent to which data move across space and time.

Data models are the result of the efforts made to structure and visualise data, so that the data can be used as representations of a specific aspect of the world.

Data provenance refers to the conditions under which data were generated.

Data science is a newly emerged research domain that includes several types of expertise relevant to data analysis, such as computational and statistical skills; epistemological expertise, including an understanding of where data fit in the processes of knowledge production; and expertise on data governance and ethics.

Data subject is a label for a person who can be identified directly or indirectly by an identifier such as name, location data, online identifier or by facets of one's identity, be they physical, physiological, genetic, mental, economic, cultural or social.

Data visualisation is the process of ordering and interacting with data in visual form. Most data visualisations aim to facilitate the discovery of patterns.

Data workers are individuals who are in a position to take decisions concerning what data should be gathered and used, for which purposes, and in which ways.

Ethics consists of philosophical reflection on what it means to be a good person.

Evidence is the use of data to provide reasons to believe in a particular claim.

FAIR is an acronym that stands for four principles introduced in 2016 as guidance for data management and sharing: Findability, Accessibility, Interoperability and Reuse.

GDPR (General Data Protection Regulation) is a piece of European legislation introduced in 2018 to protect individuals from abuse of their personal data and encourage the development of sophisticated and responsible ways of collecting, archiving, mobilising and reusing personal data.

Information society denotes a society where information is central to the capitalist system of production, innovation and consumption.

Knowledge commons is a label used to describe knowledge as a public good that contributes key insights on human life and that should be accessible without restrictions.

Knowledge society refers to a society that generates, processes, shares and makes knowledge that may be used to improve the human condition available to all its members.

Machine learning is a branch of artificial intelligence (AI) where a dataset is used by a computer to build and/or further refine a computational approach to solving a specific problem, such as image recognition or classifying information.

Metadata are structured data used to describe the characteristics of a given dataset, such as its provenance or significance.

Metrics are measures used for assessment.

Morality is the systems of norms and rules that tell us what is right and what is wrong, that is, how we should behave.

Networks are systems of interconnected things, processes or individuals. Digital networks linking computing devices are central to the transmission of digital content.

Open Science is a movement committed to promoting collaborative research practices and the widespread distribution and reuse of research components including data and models.

Platforms are programmable infrastructures upon which other software can be developed and run.

Raw data are data that have just been generated and have not been further processed.

Space of agency (for data workers) is the space within which data workers can make decisions and take responsibility for the implications of those decisions.

Statistical models are precise and concise mathematical descriptions of datasets that enable calculations.

REFERENCES

Abrieu, R., Rapetti, M., Aneja, U. and Chetty, K. (2019). How to promote worker wellbeing in the platform economy in the Global South. *G20 Insights*, Japan. Think 20 Engagement Group.

Acker, A. (2018). *Data Craft: The Manipulation of Social Media Metadata*. New York: Data & Society Research Institute.

Ackoff, R. (1989). From data to wisdom. *Journal of Applied Systems Analysis*, 16: 3–9.

Alaimo, C. and Kallinikos, J. (2017). Computing the everyday: Social media as data platforms. *The Information Society*, 33(4): 175–191.

Amano, T. and Sutherland, W. J. (2013). Four barriers to the global understanding of biodiversity conservation: wealth, language, geographical location and security. *Proceedings of the Royal Society B: Biological Sciences*, 280(1756): 20122649.

Amit-Danhi, E. R. and Shifman L. (2018). Digital political infographics: A rhetorical palette of an emergent genre. *New Media & Society*, 20(10): 3540–3559.

Amoore, L. (2009). Lines of sight: On the visualization of unknown futures. *Citizenship Studies*, 13(1): 17–30.

Amoore, L. (2011). Data derivatives: On the emergence of a security risk calculus for our times. *Theory, Culture & Society*, 28(6): 24–43.

Amoore, L. (2013). *The Politics of Possibility: Risk and Security beyond Probability*, Durham, NC: Duke University Press.

Amoore, L. (2020). *Cloud Ethics: Algorithms and the Attributes of Ourselves and Others*. Durham, NC: Duke University Press.

Anderson, C. (2008). The end of theory: The data deluge makes the scientific method obsolete. *Wired*, 23 June.

Ankeny, R. A. and Leonelli, S. (2016). Repertoires: A post-Kuhnian perspective on collaborative research. *Studies in History and Philosophy of Science Part A*, 60: 18–28.

Aouragh, M., Gürses, S., Pritchard, H. and Snelting, F. (2020). The extractive infrastructures of contact tracing apps. *Journal of Environmental Media*, 1(2): 9–1.

Aronova, E., Baker, K. S. and Oreskes, N. (2010). Big Science and Big Data in biology: From the International Geophysical Year through the International Biological Program to the Long-Term Ecological Research (LTER) Network, 1957 – Present. *Historical Studies in the Natural Sciences*, 40(2): 183–224.

Arora, P. and Rangaswamy, N. (2013). Digital leisure for development: Reframing new media practice in the global South. *Media, Culture & Society*, 35(7): 898–905.

Barry, A. and Born, G. (eds) (2013). *Interdisciplinarity: Reconfigurations of the Social and Natural Sciences*. London; New York: Routledge.

Bates, J., Cameron, D., Checco, A., Clough, P., Hopfgartner, F., Mazumdar, S., Sbaffi, L., Stordy, P. and de la Vega de León, A. (2020, January). Integrating FATE/critical data studies into data science curricula: Where Are we going and how do we get there? In *Proceedings of the 2020 Conference on Fairness, Accountability, and Transparency*. New York: Association for Computing Machinery. pp.425–435.

Bates, J., Lin, Y. W. and Goodale, P. (2016). Data journeys: Capturing the socio-material constitution of data objects and flows. *Big Data and Society*, 3: 2053951716654502.

Bauer, S. (2008). Mining data, gathering variables and recombining information: The Flexible architecture of epidemiological studies. *Studies in History and Philosophy of Science Part C: Studies in History and Philosophy of Biological and Biomedical Sciences*, 39(4): 415–428.

Bauer, G. R., Braimoh, J., Scheim, A. I. and Dharma, C. (2017). Transgender-inclusive measures of sex/gender for population surveys: Mixed-methods evaluation and recommendations. *PLoS ONE*, 12(5).

Beaulieu, A. (2001). Voxels in the brain: Neuroscience, informatics and changing notions of objectivity. *Social Studies of Science*, 31(5): 635–680.

Beaulieu, A. (2002). Images are not the (only) truth: Brain mapping, visual knowledge, and iconoclasm. *Science, Technology & Human Values*, 27(1): 53–86.

Beaulieu, A. (2004). From brainbank to database: The informational turn in the study of the brain. *Studies in History and Philosophy of Science Part C: Studies in History and Philosophy of Biological and Biomedical Sciences*, 35(2): 367–390.

Beaulieu, A. (2021). Data practices and SDGs: Organising knowledge for sustainable futures. In M. Hojer Bruun, D. Brogaard Kristensen, R. Douglas-Jones, C. Hasse, K. Høyer, B. R. Winthereik and A. Wahlberg (eds), *Handbook for the Anthropology of Technology*. London: Palgrave Macmillan.

Beaulieu, A. (2021). Organising knowledge for sustainable futures. In B. R. Winthereik and K. Hoeyer (eds), *Handbook for the Anthropology of Technology*. London: Palgrave Macmillan.

Beaulieu, A. and Estalella, A. (2012). Rethinking research ethics for mediated settings. *Information, Communication & Society*, 15(1): 23–42.

Beaulieu, A., de Rijcke, S., van Heur, B., Wouters, P., Scharnhorst, A. and Wyatt, S. (2013). Groningen Energy & Sustainability Programme, and Energy and Sustainability Research Institute Groningen. Authority and expertise in new sites of knowledge production. In *Virtual Knowledge.*, Cambridge, MA: MIT Press. pp.25–56.

Beer, D. (2015). Productive measures: Culture and measurement in the context of everyday neoliberalism. *Big Data & Society*, 2(1): 2053951715578951.

Beer, D. (2016). *Metric Power*. London: Palgrave Macmillan.

Beer, D. (2018). *The Data Gaze: Capitalism, Power and Perception*. London: Sage.

Bell, D. (1979). The social framework of the information society. In M. Dertoozos and J. Moses (eds), *The Computer Age: A Twenty-Year View*. Cambridge, MA: MIT Press. pp.500–549.

Berman, F., Rutenbar, R., Hailpern, B., Christensen, H., Davidson, S., Estrin, D., Franklin, M., Martonosi, M., Raghavan, P., Stodden, V. and Szalay, A. S. (2018). Realizing the potential of data science. *Communications of the ACM*, 61(4): 67–72.

Bezuidenhout, L. and Ratti, E. (2020). What does it mean to embed ethics in data science? An integrative approach based on the microethics and virtues. *AI & Society*, 1–15.

Bezuidenhout, L., Leonelli, S., Kelly, A. and Rappert, B. (2017). Beyond the digital divide: Towards a situated approach to open data. *Science and Public Policy*, 44(4): 464–475.

Bigo, D., Isin, E. and Ruppert, E. (eds) (2019). *Data Politics: Worlds, Subjects, Rights*. London and New York: Routledge.

Birch, K. and Muniesa, F. (eds) (2020). *Assetization: Turning Things into Assets in Technoscientific Capitalism*. Cambridge, MA: MIT Press.

Birhane, A. and Cummins, F. (2019). Algorithmic injustices: Towards a relational ethics. *Pre-print*. https://arxiv.org/abs/1912.07376 (accessed 21 February 2021).

Blasimme, A., and Vayena, E. (2020). What's next for COVID-19 apps? Governance and oversight. *Science*, 370(6518): 760–62.

Blei, D. M. and Smyth, P. (2017). Science and data science. *Proceedings of the National Academy of Sciences*, 114(33): 8689–8692.

Blondel, V. D., Esch, M., Chan, C., Clérot, F., Deville, P., Huens, E., Morlot, F., Smoreda, Z. and Ziemlicki, C. (2012). Data for development: The D4D challenge on mobile phone data. *Computers and Society [cs.CY] ArXiv*, 1210.0137.

Bogen, J. (2010). Noise in the world. *Philosophy of Science*, 77(5): 778–791.

Bonnin, N., van Andel, A. C., Kerby, J. T., Piel, A. K., Pintea, L. and Wich, S. A. (2018). Assessment of chimpanzee nest detectability in drone-acquired images. *Drones*, 2(2): 17.

Borgman, C. L. (2015). *Big Data, Little Data, No Data*. Cambridge, MA: MIT Press.

Boulton, G., Campbell, P., Collins, B., Elias, P., Hall, W., Laurie, G., O'Neill, O., Rawlins, M., Thornton, J., Vallance, P. and Walport, M. (2012). Science as an open enterprise. London: The Royal Society. https://royalsociety.org/topics-policy/projects/science-public-enterprise/report (accessed 27 April 2021).

Boyd, D. and Crawford, K. (2012). Critical questions for big data. *Information, Communication & Society*, 15(5): 662–679.

Brayne, S. (2020). *Predict and Surveil: Data, Discretion, and the Future of Policing*. New York: Oxford University Press.

Bringsjord, S. and Govindarajulu, N. S. (2018). *The Stanford Encyclopedia of Philosophy: Artificial Intelligence*. Stanford, CA: Stanford University.

British Academy and Royal Society (2017). Data management and use: Governance in the 21st century. A joint report of the Royal Society and the British Academy. https://www.thebritishacademy.ac.uk/publications/data-ai-management-use-governance-21st-century (accessed 15 February 2021).

Brown, L.X.Z., Richardson, M., Shetty, R., Crawford, A. and Hoagland, T. (2020). Report: Challenging the use of algorithm-driven decision-making in benefits determinations affecting people with disabilities. Georgetown, DC: Centre for Democracy and Technology.https://cdt.org/insights/report-challenging-the-use-of-algorithm-driven-decision-making-in-benefits-determinations-affecting-people-with-disabilities (accessed 24 February 2021).

Bruns, A. and Burgess, J. (2016). Methodological innovation in precarious spaces: The case of Twitter. In H. Snee, C. Hine, Y. Morey, S. Roberts and H. Watson (eds), *Digital Methods for Social Science: An Interdisciplinary Guide to Research Innovation*. London: Palgrave Macmillan. pp.17–33.

Brunton, F. and Nissenbaum, H. (2015). *Obfuscation: A User's Guide for Privacy and Protest*. Cambridge, MA: MIT Press.

Buolamwini, J. and Gebru, T. (2018). Gender shades: Intersectional accuracy disparities in commercial gender classification. In Conference on Fairness, Accountability and Transparency. PMLR. pp.77–91.

Burgelman, J. C., Pascu, C., Szkuta, K., Von Schomberg, R., Karalopoulos, A., Repanas, K. and Schouppe, M. (2019). Open Science, Open Data, and Open Scholarship: European policies to make science fit for the twenty-first century. *Frontiers in Big Data*, 2(43): 1–6.

Burton, P. R., Murtagh, M. J., Boyd, A., Williams, J. B., Dove, E. S., Wallace, S. E., Tasse, A. M., Little, J., Chisholm, R. L., Gaye, A. and Hveem, K. (2015). Data safe havens in health research and healthcare. *Bioinformatics*, 31(20): 3241–3248.

Cai, L. and Zhu, Y. (2015). The challenges of data quality and data quality assessment in the Big Data era. *Data Science Journal*, 14(0): 2.

Cartwright, N. and Hardie, J. (2013). *Evidence-Based Policy: A Practical Guide to Doing it Better*. Oxford, MA: Oxford University Press.

Castelfranchi, C. (2007). Six critical remarks on science and the construction of the knowledge society. *Journal of Science Communication*, 6(4): C03.

Castells, M. (1996). *The Rise of the Network Society*. Cambridge, MA: Blackwell Publishers.

Chun, W. H. K. (2016). *Updating to Remain the Same: Habitual New Media*. Cambridge, MA: MIT Press.

Cinnamon, J. (2019). Data inequalities and why they matter for development. *Information Technology for Development*, 26(2): 214–233.

Coeckelbergh, M. (2021). AI for climate: Freedom, justice, and other ethical and political challenges. *AI and Ethics*, 1(1): 67–72.

Conway, D. (2015). The data science venn diagram. *Drew Conway*. 2010. http://drewconway.com/zia/2013/3/26/the-data-science-venn-diagram (accessed 15 February 2021).

Couldry, N. and Mejias, U. A. (2019). *The Costs of Connection: How Data Are Colonizing Human Life and Appropriating It for Capitalism*. Stanford, CA: Stanford University Press.

Crawford, K. (2021). *Atlas of AI: Power, Politics, and the Planetary Costs of Artificial Intelligence*. New Haven, CT: Yale University Press.

D'Ignazio, C. and Klein, L. F. (2020). *Data Feminism*. Cambridge, MA: MIT Press.

Daston, L. (2017). *Science in the Archives: Pasts, Presents, Futures*. Chicago, IL: University of Chicago Press.

Daston, L. and Galison, P. (2007). *Objectivity*. New York: Zone Books.

de Chadarevian, S. (2018). Things and data in recent biology. *Historical Studies in the Natural Sciences*, 48(5): 648–658.

Decuyper, A., Browet, A., Traag, V., Blondel, V. D. and Delvenne, J. C. (2016). Clean up or mess up: The effect of sampling biases on measurements of degree distributions in mobile phone datasets. *ArXiv:1609.09413 [Physics]*, September.

de Rijcke, S. and Beaulieu, A. (2011). Image as interface: Consequences for users of museum knowledge. *Library Trends*, 59(4): 663–685.

de Rijcke, S. and Beaulieu, A. (2014). Networked neuroscience: Brain scans and visual knowing at the intersection of atlases and databases. In C. Coopmans, S. Woolgar, J. Vertesi and M. Lynch (eds), *Representation in Scientific Practice Revisited*. Cambridge, MA: MIT Press.

de Rijcke, S., Wouters, P. F., Rushforth, A. D., Franssen, T. P. and Hammarfelt, B. (2016). Evaluation practices and effects of indicator use – a literature review. *Research Evaluation*, 25(2).

Derksen, M. and Beaulieu, A. (2011). Social technology. In I. C. Jarvie and J. Zamora-Bonilla (eds), *The Handbook of Philosophy of Social Science*. London: Sage. pp.703–719.

Desrosières, A. (2010). *La Politique des Grands Nombres*. Paris: La Découverte.

De Veaux, R. D., Agarwal, M., Averett, M., Baumer, B. S., Bray, A., Bressoud, T. C., Bryant, L., Cheng, L. Z., Francis, A., Gould, R. and Kim, A. Y. (2017). Curriculum guidelines for undergraduate programs in data science. *Annual Review of Statistics and Its Application*, 4: 15–30.

Dormans, S. and Kok, J. (2010). An alternative approach to large historical databases: Exploring best practices with collaboratories. *Historical Methods*, 43(3): 97–107.

Drahokoupil, J. and Jepsen, M. (2017). The digital economy and its implications for labour: 1. The platform economy. *Transfer: European Review of Labour and Research*, 23(2): 103–119.

Drucker, J. (2014). *Graphesis: Visual Forms of Knowledge Production*. Cambridge, MA: Harvard University Press.

Dumit, J. and Nafus. D. (2018). The other ninety per cent: Thinking with data science, creating data studies – Joseph Dumit interviewed by Dawn Nafus. In H. Knox and D. Nafus (eds), *Ethnography for a Data-saturated World*. Manchester: Manchester University Press. pp.252–274.

Dye, C. (2014). After 2015: Infectious diseases in a new era of health and development. *Philosophical Transactions of the Royal Society B: Biological Sciences*, 369(1645): 20130426.

Ebeling, M. (2016). *Healthcare and Big Data: Digital Spectres and Phantom Objects*. London, New York: Palgrave Macmillan.

Edelman, B., Luca, M. and Svirsky, D. (2017). Racial discrimination in the sharing economy: Evidence from a field experiment. *American Economic Journal: Applied Economics*, 9(2): 1–22.

Edwards, P. N. (2010). *A Vast Machine: Computer Models, Climate Data, and the Politics of Global Warming*. Cambridge, MA: MIT Press.

Edwards, P. N. (2019). Knowledge infrastructures under siege: Climate data as memory, truce, and target. In D. Bigo, E. Isin and E. Ruppert (eds), *Data Politics: Worlds, Subjects, Rights*. London: Routledge. pp.21–42.

Edwards, P. N., Mayernik, M. S., Batcheller, A. L., Bowker, G. C. and Borgman, C. L. (2011). Science friction: Data, metadata, and collaboration. *Social Studies of Science*, 41(5): 667–690.

Erikson, S. L. (2018). Cell phones ≠ self and other problems with big data detection and containment during epidemics. *Medical Anthropology Quarterly*, 32(3): 315–339.

Eubanks, V. (2018). *Automating Inequality: How High-Tech Tools Profile, Police, and Punish the Poor*. Illustrated edition. New York: St Martin's Press.

European Commission (2016). *Open Innovation, Open Science, Open to the World. A Vision for Europe*. Luxembourg: Publications Office of the European Union.

Eyert, F., Irgmaier, F. and Ulbricht, L. (2020). Extending the framework of algorithmic regulation. The Uber case. *Regulation & Governance*, https://doi.org/10.1111/rego.12371 (accessed 21 February 2021).

Featherstone, R. (2014). Visual research data: An infographics primer. *The Journal of the Canadian Health Libraries Association. [Journal de l'Association des Bibliothèques de la Santé du Canada]*, 35(3): 147–150.

Fecher, B. and Friesike, S. (2014). Open Science: One term, five schools of thought. *Opening Science*, 17.

Finzer, W. (2013). The data science education dilemma. *Technology Innovations in Statistics Education*, 7(2).

Fleming, L., Tempini, N., Gordon-Brown, H., Nichols, G. L., Sarran, C., Vineis, P., Leonardi, G., et al. (2017). Big data in environment and human health. *Oxford Research Encyclopedia of Environmental Science*. Oxford: Oxford University Press.

Floridi, L. (2011). *The Philosophy of Information*. Oxford: Oxford University Press.

Floridi, L. (2013). Distributed morality in an information society. *Science and Engineering Ethics*, 19(3): 727–743.

Floridi, L. (2014). *The Fourth Revolution: How the Infosphere is Reshaping Human Reality*. Oxford: Oxford University Press.

Floridi, L. (2015). *The Ethics of Information*. Oxford: Oxford University Press.

Floridi, L. (2017). Robots, jobs, taxes, and responsibilities. *Philosophy & Technology*, 1(30): 1–4.

Floridi, L. and Illari, P. (eds) (2014). *The Philosophy of Information Quality*. Synthese Library. Basel: Springer International Publishing.

Floridi, L., Cowls, J., Beltrametti, M., Chatila, R., Chazerand, P., Dignum, V., Luetge, C., Madelin, R., Pagallo, U., Rossi, F. and Schafer, B. (2018). AI4People – an ethical framework for a good AI society: Opportunities, risks, principles, and recommendations. *Minds and Machines*, 28(4): 689–707.

Francois, K., Monteiro, C. and Allo, P. (2020). Big-data literacy as a new vocation for statistical literacy. *Statistics Education Research Journal*, 19(1).

Freeth, R. and Caniglia, G. (2020). Learning to collaborate while collaborating: advancing interdisciplinary sustainability research. *Sustainability Science*, 15(1): 247–261.

Fricker, M. (2009). *Epistemic Injustice: Power and the Ethics of Knowing*. Oxford: Oxford University Press.

Frigg, R. and Hartmann, S. (2020). Models in Science. *The Stanford Encyclopedia of Philosophy* (Spring 2020 Edition), Edward N. Zalta (ed.) https://plato.stanford.edu/archives/spr2020/entries/models-science/.

Frosh, S. (2016). Relationality in a time of surveillance: Narcissism, melancholia, paranoia. *Subjectivity*, 9(1): 1–16.

Gabriel, L. and Casemore, R. (eds) (2009). *Relational Ethics in Practice: Narratives from Counselling and Psychotherapy*. London: Routledge.

Garber, A. M. (2019). Data science: What the educated citizen needs to know. *Harvard Data Science Review*, 1(1).

Garcia, P., Sutherland, T., Cifor, M., Chan, A. S., Klein, L., D'Ignazio, C. and Salehi, N. (2020). No: Critical refusal as feminist data practice. In *Conference*

Companion Publication of the 2020 on Computer Supported Cooperative Work and Social Computing. New York: Association for Computing Machinery. pp.199–202.

Gibbons, M. (1994). *The New Production of Knowledge: The Dynamics of Science and Research in Contemporary Societies.* London: Sage.

Gillespie, T. (2010). The politics of 'platforms' – Tarleton Gillespie, 2010. *New Media & Society,* February.

Gillespie, T. (2017). The platform metaphor, revisited. *Culture Digitally* (blog). http://culturedigitally.org/2017/08/platform-metaphor (accessed 15 February 2021).

Given, L. M. (2008). *The Sage Encyclopedia of Qualitative Research Methods.* Vols 1–2. Thousand Oaks, CA: Sage.

Green, S. and Vogt, H. (2016). Personalizing medicine: disease prevention in silico and in socio. *appear in Humana. Mente Journal of Philosophical Studies,* 30.

Greene, D., Hoffmann, A. L. and Stark, L. (2019). Better, nicer, clearer, fairer: A critical assessment of the movement for ethical artificial intelligence and machine learning. In *Proceedings of the 52nd Hawaii International Conference on System Sciences.*

Greenberg, J. (2017). Big metadata, smart metadata, and metadata capital: Toward greater synergy between data science and metadata. *Journal of Data and Information Science,* 2(3): 19–36.

GSMA (2018). Helping end tuberculosis in India by 2025. https://www.gsma.com/betterfuture/wp-content/uploads/2018/12/Helping_end_Tuberculosis_in_India_by_2025.pdf (accessed 27 April 2021).

Gstrein, O. J., Bunnik, A. and Zwitter, A. (2019). Ethical, legal and social challenges of predictive policing. *Católica Law Review, Direito Penal,* 3(3): 77–98.

Harris, A., Kelly, S. and Wyatt, S. (2016). *CyberGenetics: Health Genetics and New Media.* London and New York: Routledge.

Hazard, L., Cerf, M., Lamine, C., Magda, D. and Steyaert, P. (2020). A tool for reflecting on research stances to support sustainability transitions. *Nature Sustainability,* 3(2): 89–95.

Hess, C. and Ostrom, E. (2007). *Understanding Knowledge as Commons.* Cambridge, MA: MIT Press.

Hewson, M. (1999). Did global governance create informational globalism? *Approaches to Global Governance Theory,* 97–113.

Hey, T., Tansley, S. and Tolle, K. (eds) (2009). *The Fourth Paradigm: Data-Intensive Scientific Discovery.* Redmond, WA: Microsoft Research.

Hilgartner, S. (1995). Biomolecular databases: New communication regimes for biology? *Science Communication,* 17(1): 240–263.

Hilgartner, S. (2017). *Reordering Life: Knowledge and Control in the Genomics Revolution*. Cambridge, MA: MIT Press.

Hine, C. (2001). Web pages, authors and audiences: The meaning of a mouse click. *Information, Communication & Society*, 4(2): 182–198.

Hoang, L., Blank, G. and Quan-Haase, A. (2020). The winners and the losers of the platform economy: Who participates? *Information, Communication and Society*, 23(5): 681–700.

Hogle, L. F. (2016). The ethics and politics of infrastructures: Creating the conditions of possibility for big data in medicine. In *The Ethics of Biomedical Big Data*. Cham: Springer. pp.397–427.

Holstein, K., Wortman Vaughan, J., Daumé III, H., Dudik, M. and Wallach, H. (2019, May). Improving fairness in machine learning systems: What do industry practitioners need? In *Proceedings of the 2019 CHI Conference on Human Factors in Computing Systems. CHI '19*. New York: Association for Computing Machinery. pp.1–16.

Hood, L. and Friend, S. H. (2011). Predictive, personalized, preventive, participatory (P4) cancer medicine. *Nature Reviews. Clinical Oncology*, 8(3): 184–187.

Horton, N. J. and Hardin, J. S. (2015). Teaching the next generation of statistics students to 'think with data': Special issue on statistics and the undergraduate curriculum. *The American Statistician*, 69(4): 259–265.

Hristova, D., Williams, M. J., Musolesi, M., Panzarasa, P. and Mascolo, C. (2016). Measuring urban social diversity using interconnected geo-social networks. In *Proceedings of the 25th International Conference on World Wide Web. WWW '16*. Montréal: International World Wide Web Conferences Steering Committee. pp.21–30.

International Telecommunications Union (2018). Measuring the information society report, 1(1). Geneva: International Telecommunications Union. p.204.

Irwin, A. (2018). No PhDs needed: How citizen science is transforming research. *Nature*, 562(7728): 480–482.

Jasanoff, S. (2003). Technologies of humility: Citizen participation in governing science. *Minerva*, 41(3): 223–244.

Jemielniak, D. and Przegalinska, A. (2020). *Collaborative Society*. Cambridge, MA: MIT Press.

Jones, K. H., Ford, D. V., Jones, C., Dsilva, R., Thompson, S., Brooks, C. J., Heaven, M. L., Thayer, D. S., McNerney, C. L. and Lyons, R. A. (2014). A Case Study of the Secure Anonymous Information Linkage (SAIL) Gateway: A Privacy-Protecting Remote Access System for Health-Related Research and Evaluation. *Journal of Biomedical Informatics, Special Issue on Informatics Methods in Medical Privacy*, 50 (August): 196–204.

Jones, K. M., Ankeny, R. A. and Cook-Deegan, R. (2018). The Bermuda Triangle: The pragmatics, policies, and principles for data sharing in the history of the human Genome Project. *Journal of the History of Biology*, 51(4): 693–805.

Kaldestad, Ø. H. (2016). *250,000 words of app terms and conditions*. https://www.forbrukerradet.no/side/250000-words-of-app-terms-and-conditions (accessed 21 February 2021).

Kallinikos, J. and Tempini, N. (2014). Patient data as medical facts: Social media practices as a foundation for medical knowledge creation. *Information Systems Research*, 25(4): 817–833.

Kaye, J., Whitley, E. A., Lund, D., Morrison, M., Teare, H. and Melham, K. (2015). Dynamic consent: A patient interface for twenty-first century research networks. *European Journal of Human Genetics*, 23(2): 141–146.

Kennedy, H., Hill, R. L., Aiello, G. and Allen, W. (2016). The work that visualisation conventions do. *Information, Communication & Society*, 19(6): 715–735.

Kitchin, R. (2021). *The Data Revolution: Big Data, Open Data, Data Infrastructures and Their Consequences*. London: Sage.

Kitchin, R. and Dodge, M. (2007). Rethinking maps. *Progress in Human Geography*, 31(3): 331–344.

Kitchin, R. and Dodge, M. (2011). *Code/Space: Software and Everyday Life*. Cambridge, MA: MIT Press.

Kitchin, R. and McArdle, G. (2016). What makes big data, big data? Exploring the ontological characteristics of 26 datasets. *Big Data & Society*, 3(1): 2053951716631130.

Kolaczyk, E., Lee, M. M., Liu, J. and Parker, M. S. (2021). We need a (responsible!) data science rapid response network. *Harvard Data Science Review*. https://doi.org/10.1162/99608f92.2794e78d

Krige, J. and Leonelli, S. (2021). Mobilizing the translational history of knowledge flows: COVID-19 and the politics of knowledge at the borders. *History and Technology*. https://doi.org/10.1080/07341512.2021.1890524

Kuner, C. (2017). Reality and illusion in EU data transfer regulation post schrems. *German Law Journal*, 18(4): 881–918.

Lagoze, C. (2014). Big Data, data integrity, and the fracturing of the control zone. *Big Data & Society*, 1(2): 2053951714558281.

Lane, J. (2020). *Democratising our Data: A Manifesto*. Cambridge, MA: MIT Press.

Lave, J. and Wenger, E. (1991). *Situated Learning: Legitimate Peripheral Participation. Learning in Doing: Social, Cognitive and Computational Perspectives*. Cambridge: Cambridge University Press.

Leonelli, S. (2013). Why the current insistence on open access to scientific data? Big Data, knowledge production, and the political economy of contemporary biology. *Bulletin of Science, Technology & Society*, 33(1): 6–11.

Leonelli, S. (2016a). *Data-Centric Biology: A Philosophical Study*. Chicago, IL: University of Chicago Press.

Leonelli, S. (2016b). Locating ethics in data science: Responsibility and accountability in global and distributed knowledge production. *Philosophical Transactions of the Royal Society: Part A*, 374: 20160122.

Leonelli, S. (2017). Global data quality assessment and the situated nature of 'best' research practices in biology. *Data Science Journal*, 16.

Leonelli, S. (2018). Rethinking reproducibility as a criterion for research quality. In *Including a Symposium on Mary Morgan: Curiosity, Imagination, and Surprise*. Bingley: Emerald Publishing Limited.

Leonelli, S. (2019). Data governance is key to interpretation: Reconceptualizing data in data science. *Harvard Data Science Review*.

Leonelli, S. (2020). Big data and scientific research. *Stanford Encyclopedia for Philosophy*. Stanford, CA: Stanford University.

Leonelli, S. (2021). Data science in times of pan(dem)ic. *Harvard Data Science Review*, 3(1). https://doi.org/10.1162/99608f92.fbb1bdd6

Leonelli, S. and Tempini, N. (2018). Where health and environment meet: The use of invariant parameters in big data analysis. *Synthese*, https://doi.org/10.1007/s11229-018-1844-2

Leonelli, S. and Tempini, N. (2020). *Data Journeys in the Sciences*. Basel: Springer International.

Leonelli, S., Lovell, B., Fleming, L., Wheeler, B. and Williams, H. (2021). From FAIR data to fair data use: Methodological data fairness in health-related social media research. *Big Data and Society*, 8(1). https://doi.org/10.1177/20539517211010310

Leslie, D. (2019). Understanding artificial intelligence ethics and safety: A guide for the responsible design and implementation of AI systems in the public sector. *The Alan Turing Institute, e-Prints*. https://ui.adsabs.harvard.edu/abs/2019arXiv190605684L (accessed 21 February 2021).

Leslie, D. (2020). Tackling COVID-19 through responsible AI innovation: Five steps in the right direction. *Harvard Data Science Review*, Special Issue 1-COVID-19 (June). https://hdsr.mitpress.mit.edu/pub/as1p81um/release/3 (accessed 5 May 2021).

Letouzé, E. (2015). *Big Data & development: An overview*. Data-Pop Alliance White Paper Series. Data-Pop Alliance, World Bank Group, Harvard Humanitarian Initiative.

Li, C. and Sugimoto, S. (2017). Provenance description of metadata vocabularies for the long-term maintenance of metadata. *Journal of Data and Information Science*, 2(2): 41–55.

Lindstrom, M. (2016). *Small Data: The Tiny Clues that Uncover New Worlds.* St Martin's Press.

Lowrie, I. (2017). Algorithmic rationality: Epistemology and efficiency in the data sciences. *Big Data & Society*, 4(1): 1–13.

Lucivero, F. (2020). Big data, big waste? A reflection on the environmental sustainability of Big Data initiatives. *Science and Engineering Ethics*, 26(2): 1009–1030.

Lucivero, F. and Prainsack, B. (2015). The life stylisation of healthcare? 'Consumer genomics' and mobile health as technologies for healthy lifestyle. *Applied & Translational Genomics*, 4: 44–49.

Lupton, D. and Michael, M. (2017). 'For me, the biggest benefit is being ahead of the game': The use of social media in health work. *Social Media + Society*, 3(2).

Mackenzie, A. (2017). *Machine Learners: Archaeology of a Data Practice.* Cambridge, MA: MIT Press.

Magalhães, J. C. and Couldry, N. (2020). Tech giants are using this crisis to colonize the welfare system. Jacobin. https://jacobinmag.com/2020/04/tech-giants-coronavirus-pandemic-welfare-surveillance (accessed 21 February 2021).

Marsh, S. (2019). One in three councils using algorithms to make welfare decisions. *The Guardian*, 15 October 2019. www.theguardian.com/society/2019/oct/15/councils-using-algorithms-make-welfare-decisions-benefits (accessed 15 February 2021).

Mao, Y., Wang, D., Muller, M., Varshney, K. R., Baldini, I., Dugan, C. and Mojsilović, A. (2019). How data scientists work together with domain experts in scientific collaborations: To find the right answer or to ask the right question? *Proceedings of the ACM on Human-Computer Interaction*, 3(1): 1–23.

Mauthner, N. S. and Doucet, A. (2008). 'Knowledge Once divided can be hard to put together again': An epistemological critique of collaborative and team-based research practices. *Sociology*, 42(5): 971–985.

Mayer-Schönberger, V. and Cukier, K. (2013). *Big Data: A Revolution That Will Transform How We Live, Work, and Think.* Boston, MA: Houghton Mifflin Harcourt.

Mayernik, M. S. (2019). Metadata accounts: Achieving data and evidence in scientific research. *Social Studies of Science*, 49(5): 732–757.

Mayo, D. (2018). *Statistical Inference as Severe Testing: How to Get Beyond the Statistics Wars.* Cambridge: Cambridge University Press.

Mayo, D. G. and Spanos, A. (eds) (2009). *Error and Inference.* Cambridge: Cambridge University Press.

Mayo, D. G. and Spanos, A. (2011). Error statistics. In P. S. Bandyopadhyay and M. R. Forster (eds), *Philosophy of Statistics*. Vol. 7. Amsterdam: Elsevier. pp.153–198.

Medina-Perea, I. A., Bates, J. and Cox, A. (2019). Using data journeys to inform research design: Socio-cultural dynamics of patient data flows in the UK healthcare sector. In *iConference 2019 Proceedings*.

Meng, X.-L. (2019). Data science: An artificial ecosystem. *Harvard Data Science Review*, 1(1). https://doi.org/10.1162/99608f92.ba20f92

Merelli, I., Pérez-Sánchez, H., Gesing, S. and D'Agostino, D. (2014). Managing, analysing, and integrating Big Data in medical bioinformatics: Open problems and future perspectives. *BioMed Research International*.

Mikroyannidis, A., Domingue, J., Phethean, C., Beeston, G. and Simperl, E. (2017). The European Data Science Academy: Bridging the data science skills gap with open courseware. In Open Education Global Conference 2017, 8–10 March, Cape Town, South Africa.

Mikroyannidis, A., Domingue, J., Phethean, C., Beeston, G. and Simperl, E. (2018). Designing and delivering a curriculum for data science education across Europe. In M. E. Auer, D. Guralnick and I. Simonics (eds), *Teaching and Learning in a Digital World. Advances in Intelligent Systems and Computing*. Cham: Springer International Publishing. pp.540–550.

Mirowski, P. (2018). The future(s) of open science. *Social Studies of Science*, 48(2): 171–203.

Mitchell, S., Potash, E., Barocas, S., D'Amour, A. and Lum, K. (2021). Algorithmic fairness: Choices, assumptions, and definitions. *Annual Review of Statistics and Its Application*, 8(1).

Mittelstadt, B. (2019). Principles alone cannot guarantee ethical AI. *Nature Machine Intelligence*, 1(11): 501–507.

Mittelstadt, B. D., Allo, P., Taddeo, M., Wachter, S. and Floridi, L. (2016). The ethics of algorithms: Mapping the debate. *Big Data & Society*, 3(2): 2053951716679679.

Morrison, M. (2014). *Reconstructing Reality: Models, Mathematics, and Simulations*. Oxford: Oxford University Press.

Moss, E. and Metcalf, J. (2020). *Ethics Owners: A New Model of Organizational Responsibility in Data-Driven Technology Companies*. New York: Data and Society Research Institute.

Noble, S. U. (2018). *Algorithms of Oppression: How Search Engines Reinforce Racism*. Illustrated edition. New York: NYU Press.

Nowotny, H., Scott, P. and Gibbons, M. (2001). *Re-Thinking Science: Knowledge and the Public in an Age of Uncertainty*. Cambridge: Polity Press.

Nuffield Council on Bioethics (2015). Chapter 3: Public Interest. In The collection, linking and use of data in biomedical research and health care: Ethical issues. London: Nuffield Council on Bioethics. pp.53–55. https://www.nuffieldbioethics.org/publications/biological-and-health-data (accessed 15 February 2021).

O'Malley, M. A. and Soyer, O. S. (2012). The roles of integration in molecular systems biology. *Studies in History and Philosophy of Science Part C: Studies in History and Philosophy of Biological and Biomedical Sciences*, 43(1): 58–68.

O'Neill, C. (2016). *Weapons of Math Destruction: How Big Data Increases Inequality and Threatens Democracy*. New York: Crown Archetype.

O'Neill, C. and Schutt, R. (2013). *Doing Data Science: Straight Talk from the Frontline*. Beijing, Sebastopol, CA: O'Reilly. pp.1–50.

Organisation for Economic Co-operation and Development (OECD) (2014). OECD science, technology and industry outlook 2014. Paris: OECD. https://doi.org/10.1787/19991428 (accessed 21 February 2021).

Organisation for Economic Co-operation and Development (OECD) (2017). Recommendations of the Council on Health Data Governance. Paris: OECD. OECD/LEGAL/0433. www.oecd.org/els/health-systems/health-data-governance.htm (accessed 15 February 2021).

Palmer, C. L., Thomer, A. K., Baker, K. S., Wickett, K. M., Hendrix, C. L., Rodman, A., Sigler, S. and Fouke, B. W. (2017). Site-based data curation based on hot spring geobiology. *PLoS ONE*, 12(3): p.e0172090.

Palmer, S. (2015). *Data Science for the C-Suite*. New York: Digital Living Press.

Pasquale, F. (2015). *The Black Box Society: The Secret Algorithms That Control Money and Information*. Cambridge, MA: Harvard University Press.

Pickering, A. (2011). *The Cybernetic Brain: Sketches of Another Future*. Chicago, IL: University of Chicago Press.

Pitcan, M., Marwick, A. E. and Boyd, D. (2018). Performing a vanilla self: respectability politics, social class, and the digital world. *Journal of Computer-Mediated Communication*, 23(3): 163–179.

Plantin, J. C., Lagoze, C., Edwards, P. N. and Sandvig, C. (2018). Infrastructure studies meet platform studies in the age of Google and Facebook. *New Media & Society*, 20(1): 293–310.

Poell, T., Nieborg, D. and van Dijck, J. (2019). Platformisation. *Internet Policy Review*, 8(4): 1–13.

Porter, T. M. (1995). *Trust in Numbers: The Pursuit of Objectivity in Science and Public Life*. Princeton, NJ: Princeton University Press.

Porter, T. M. and de Chadarevian, S. (2018). Introduction: Scrutinizing the data world. *Historical Studies in the Natural Sciences*, 48(5): 549–556.

Postigo, H. and O'Donnell, C. (2016). The sociotechnical architecture of information networks. In *The Handbook of Science and Technology Studies*. Cambridge, MA: MIT Press, pp.583–608.

Prainsack, B. and Buyx, A. (2017). *Solidarity in Biomedicine and Beyond*. Cambridge: Cambridge University Press.

Prey, R. (2020). In D. Stark (ed.), *The Performance Complex: Competition and Competitions in Social Life*. Oxford: Oxford University Press. pp. 241–25

Price, D. J. (1963). *Little Science, Big Science*. New York: Columbia University Press.

Radder, H. (2009). The philosophy of scientific experimentation: A review. *Automated Experimentation*, 1(1): 2.

Rangaswamy, N. and Arora, P. (2016). The mobile internet in the wild and every day: Digital leisure in the slums of urban India. *International Journal of Cultural Studies*, 19(6): 611–626.

Rappert, B. and Selgelid, M. J. (2013). *On the Dual Uses of Science and Ethics Principles, Practices, and Prospects*. Canberra: ANU Press.

Rieder, G. and Simon, J. (2016). Datatrust: Or the political quest for numerical evidence and the epistemologies of Big Data. *Big Data & Society*, 3(1): 2053951716649398.

Roussi, A. (2020). Resisting the rise of facial recognition. *Nature*, 587: 350–353. https://www.nature.com/articles/d41586-020-03188-2 (accessed 21 February 2021).

Sadowksi, J. (2019). When data is capital: Datafication, accumulation, and extraction, *Big Data & Society*, 6(1) https://doi.org/10.1177/20539 51718820549.

Sample, I. (2019). Maths and tech specialists need Hippocratic oath, says academic. *The Guardian*, 16 August. https://www.theguardian.com/science/2019/aug/16/mathematicians-need-doctor-style-hippocratic-oath-says-academic-hannah-fry (accessed 21 February 2021).

Schutt, R. and O'Neill, C. (2013). *Doing Data Science: Straight Talk from the Frontline*. Sebastopol, CA: O'Reilly Media, Incorporated.

Shapin, S. (1995). *A Social History of Truth: Civility and Science in Seventeenth-Century England*. Chicago, IL: University of Chicago Press.

Sharon, T. and Zandbergen, D. (2017). From data fetishism to quantifying selves: Self-tracking practices and the other values of data. *New Media & Society*, 19(11): 1695–1709.

Shaver, L. (2010). The right to science and culture. *Wisconsin Law Review*, 1: 121–184.

Shaw, J. and Graham, M. (2017). An informational right to the city? Code, content, control, and the urbanization of information. *Antipode*, 49(4): 907–927.

Shove, E. (2007). *The Design of Everyday Life*. New York: Berg.

Shove, E., Watson, M., Hand, M. and Ingram, J. (2007). Reproducing digital photography. *The Design of Everyday Life*. New York: Berg.

Sinnenberg, L., Buttenheim, A. M., Padrez, K., Mancheno, C., Ungar, L. H. and Merchant, R. M. (2017). Twitter as a tool for health research: A systematic review. *American Journal of Public Health*, 107(1): 1–5.

Sloane, M. and Moss, E. (2019). AI's social sciences deficit. *Nature Machine Intelligence*, 1(8): 330–331.

Srnicek, N. (2016). *Platform Capitalism*. Cambridge; Malden, MA: Polity Press.

Stark, L. and Hoffmann, A. L. (2019). Data is the new what? Popular metaphors and professional ethics in emerging data culture. *Journal of Cultural Analytics*, 1(1): 11052.

Starosielski, N. (2015). *The Undersea Network*. Durham, NC: Duke University Press.

Stephens, M. (2013). Gender and the GeoWeb: Divisions in the production of user-generated cartographic information. *GeoJournal*, 78(6): 981–996.

Sterner, B. and Franz, N. M. (2017). Taxonomy for humans or computers? Cognitive pragmatics for Big Data. *Biological Theory*, 12(2): 99–111.

Strasser, B. (2019). *Collecting Experiments: The Making of Big Data Biology*. Chicago, IL: University of Chicago Press.

Strasser, B. J. and Edwards, P. (2015). Open Access: Publishing, commerce, and the scientific ethos. SSIC report 9. Bern: Swiss Science and Innovation Council. https://citizensciences.net/wp-content/plugins/zotpress/lib/request/request.dl.php?api_user_id=424601&dlkey=BDEJXNRZ&content_type=application/pdf (accessed 15 February 2021).

Strathern, M. (2005). Anthropology and interdisciplinarity. *Arts and Humanities in Higher Education*, 4(2): 125–35.

Swan, A. and Brown, S. (2008). The Skills, role and career structure of data scientists and curators: An assessment of current practice and future needs. Key perspectives, consultants in scholarly information. Report to the JISC. Truro: JISC.

Symons, J. and Alvarado, R. (2016). Can we trust Big Data? Applying philosophy of science to software. *Big Data & Society*, 3(2): 2053951716664747.

Symons, J. and Horner, J. (2014). Software intensive science. *Philosophy & Technology*, 27(3): 461–477.

Taylor L. (2017). What is data justice? The case for connecting digital rights and freedoms globally. *Big Data & Society*, 4(2): 205395171773633.

Tempini, N. (2015). Governing PatientsLikeMe: Information production and research through an open, distributed, and data-based social media network. *The Information Society*, 31(2): 193–211.

Tempini, N. (2017). Till data do us part: Understanding data-based value creation in data-intensive infrastructures. *Information and Organization*, 27(4): 191–210.

Tempini, N. and Leonelli, S. (2018). Concealment and Discovery: The Role of Information Security in Biomedical Data Re-Use. *Social Studies of Science*, 48(5): 663–690.

Tenner, E. (2018). Who's afraid of the frightful five? Monopoly and culture in the digital age. *The Hedgehog Review*, 20(1): 68–78.

Thayyil, N. (2018). Constructing global data: Automated techniques in ecological monitoring, precaution and reification of risk. *Big Data & Society*, 5(1): 2053951718779407.

The Economist (2020). Winners from the pandemic: Big tech's Covid-19 opportunity. *The Economist*, Leaders, 4 April 2020. https://www.economist.com/leaders/2020/04/04/big-techs-covid-19-opportunity (accessed 21 February 2021).

Thrift, N. (2005). *Knowing Capitalism*. London: Sage.

Ticona, J. (2016). Phones, but no papers. Medium, points, 30 November. https://points.datasociety.net/phones-but-no-papers-e580e824ed6 (accessed 21 February 2021).

Tufte, E. (1983). *The Visual Display of Quantitative Information*. Cheshire, CT: Graphic Press.

Turnhout, E., Dewulf, A. and Hulme, M. (2016). What does policy-relevant global environmental knowledge do? The cases of climate and biodiversity. *Current Opinion in Environmental Sustainability*, 18: 65–72.

Tutton, R. (2016). Personal genomics and its sociotechnical transformations. In *Genomics and Society*. New York: Academic Press. pp.1–20.

United Nations (2018). Sustainable Development Goals report 2018. New York: UN.

United Nations (UN) News (2016). UN agency and google collaborate on satellite data tools to manage natural resources. https://news.un.org/en/story/2016/04/526802-un-agency-and-google-collaborate-satellite-data-tools-manage-natural-resources (accessed 15 February 2021).

Vallor, S. (2018). *Technology and the Virtues: A Philosophical Guide to a Future Worth Wanting*. Oxford: Oxford University Press.

Van der Vlist, F. N. (2016). Accounting for thes: Investigating commensuration and Big Data practices at Facebook. *Big Data & Society*, 3(1): 2053951716631365.

Van Dijck, J. (2020). Governing digital societies: Private platforms, public values. *Computer Law & Security Review*, 36: 105377.

Van Dijck, J., Poell, T. and De Waal, M. (2018). *The Platform Society: Public Values in a Connective World*. Oxford: Oxford University Press.

Van Doorn, N. and Badger, A. (2020). Platform capitalism's hidden abode: Producing data assets in the gig economy. *Antipode*, 52(5): 1475–1495.

Van Horn, J. D. and Toga, A. W. (2009). Is it time to re-prioritize neuroimaging databases and digital repositories? *NeuroImage*, 47(4): 1720–1734.

Vayena, E. and Prainsack, B. (2013). Regulating genomics: Time for a broader vision. *Science Translational Medicine*, 5(198): 198ed12.

Vayena, E. and Tasioulas, J. (2015). 'We the scientists': A human right to citizen science. *Philosophy & Technology*, 28(3): 479.

Vayena E, and Tasioulas J. (2016) The dynamics of big data and human rights: the case of scientific research. *Philosophical Transactions of the Royal Society of London. A* 374: 20160129. http://dx.doi.org/10.1098/rsta.2016.0129

Veale, M. and Binns, R. (2017). Fairer machine learning in the real world: Mitigating discrimination without collecting sensitive data. *Big Data & Society*, 4(2).

Vinuesa, R., Azizpour, H., Leite, I., Balaam, M., Dignum, V., Domisch, S., Felländer, A., Langhans, S. D., Tegmark, M. and Fuso, N. (2020). The role of artificial intelligence in achieving the Sustainable Development Goals. *Nature Communications*, 11(1): 233.

Von der Leyen, U. (2019). Political guidelines for the next European Commission 2019–2024. PE, 658. Brussels: European Commission.

Von Oertzen, C. (2018). Datafication and spatial visualization in nineteenth-century census statistics. *Historical Studies in the Natural Sciences*, 48(5): 568–580.

Weingart, P. and Padberg. B. (2014). University experiments in interdisciplinarity: obstacles and opportunities. *University Experiments in Interdisciplinarity*. Bielefeld: Transcript.

Wessels, B., Finn, R. L., Wadhwa, K. and Sveinsdottir, T. (2017). *Open Data and the Knowledge Society*. Amsterdam: Amsterdam University Press.

Wijmenga, C. (2019). Our opportunities lie in data. Presented at the opening of academic year 2019–20, University of Groningen, The Netherlands, 2 September. https://www.rug.nl/about-ug/latest-news/news/archief2019/nieuwsberichten/0823-speech-rector-magnificus-september-2019.pdf (accessed 15 February 2021).

Wilkinson, M. D., Dumontier, M., Aalbersberg, I. J., Appleton, G., Axton, M., Baak, A., Blomberg, N., Boiten, J. W., da Silva Santos, L. B., Bourne, P. E. and Bouwman, J. (2016). The FAIR guiding principles for scientific data management and stewardship. *Scientific Data*, 3(1): 1–9.

Wilsdon, J., Bar-ilan, J., Frodeman, R., Lex, E., Peters, I. and Wouters, P. F. (2017). Next-generation metrics: Responsible metrics and evaluation for open science. Report of the European Commission Expert Group on Altmetrics. Brussels: European Commission.

Wolff, J. and Atallah, N. (2020). Early GDPR penalties: Analysis of implementation and fines through May 2020. SSRN Paper ID 3748837. Rochester, NY: Social Science Research Network.

World Health Organization (WHO) (2019). Global tuberculosis report 2019. Geneva: World Health Organization.

Wouters, P., Beaulieu, A., Scharnhorst, A. and Wyatt, S. (eds) (2013). *Virtual Knowledge: Experimenting in the Humanities and the Social Sciences*. Cambridge, MA: MIT Press.

Yeung, K. (2018). Algorithmic regulation: A critical interrogation. *Regulation & Governance*, 12(4): 505–523.

Zhang, A. X., Muller, M. and Wang, D. (2020). How do data science workers collaborate? Roles, workflows, and tools. *ArXiv:2001.06684 [Cs, Stat]*, April.

Zook, M., Barocas, S., Crawford, K., Keller, E., Gangadharan, S. P., Goodman, A., Hollander, R., Koenig, B. A., Metcalf, J., Narayanan, A. and Nelson, A. (2017). Ten simple rules for responsible big data research. *PLoS Computational Biology*, 13(3): e1005399–e1005399.

Zuboff, S. (2019). *The Age of Surveillance Capitalism: The Fight for a Human Future at the New Frontier of Power*. New York: PublicAffairs.

Zwitter, A. (2014). Big Data ethics. *Big Data & Society*, 1(2): 2053951714559253. https://doi.org/10.1177/2053951714559253

Zwitter, A. J., Gstrein, O. J. and Yap, E. (2020). Digital identity and the blockchain: Universal identity management and the concept of the 'self-sovereign' individual. *Frontiers in Blockchain*, 3. https://doi.org/10.3389/fbloc.2020.00026

INDEX

Page numbers in *italic* indicate figures and tables, and in **bold** indicate glossary terms.